The Soul of Napoleon

I0528994

LÉON BLOY

Translated By Richard Robinson

Sunny Lou Publishing Company
Portland, Oregon, USA
http://www.sunnyloupublishing.com

2nd Edition: February 1, 2024
Original Publication Date: April 5, 2021

ISBN: 978-1-955392-55-6

* * *

This translation from French is based on the Mercure de France edition of *L'âme de Napoléon*, Paris, 1912.

Contents

Foreword

> *"I do not at all believe that Napoleon had been an evil man, and I believe even less that he had been among the damned, as affirm, with a silly emphasis, the imbecilic devotees.... of the Restoration. Paradise without my Emperor, – I do not conceive it."* – Léon Bloy

> *"A democracy is nothing more than mob rule, where fifty-one percent of the people may take away the rights of the other forty-nine."* — Thomas Jefferson

I'm a huge fan of Napoleon I, or Bonaparte. Until about a month ago, as it turns out, it wasn't my intention to translate this book, not yet. I had something else on my mind. Then I stumbled, tripped, fell over this article in the March 6, 2021 issue of the *Economist*: "Despot, genius or both? France argues about Napoléon: As his bicentenary approaches, tempers rise."

When I saw the title to the article, I got excited. Anything and everything about Napoleon, his history, his achievements, his battles, his administration, his legacy, his personal life, the ground he stood on – excite me. Always has, always will. Call me naïve or childish: I'm a fan of Napoleon.

But as I started to read the article, quickly my enthusiasm changed. True to the article's subtitle, I got incensed, and my temper rose. It rose to the degree where I nearly put down the article (well, nearly clicked away from it) when I got to this section: "The Black Lives Matter movement has emboldened those who reject any celebration of a leader who reintroduced slavery to the French West Indies in 1802. Nicholas Mayer-Rossignol, the Socialist mayor of Rouen, says he wants to replace the imposing bronze statue of the emperor on horseback that stands outside his Normandy town hall."

Now, I'm not a proponent of slavery. It's almost silly that I should feel the need to say that. Nor violence. Anyone who knows me knows I tell the truth. But it is a historical fact, slavery, and not just a recent one. To talk of slavery in the context of Napoleon seems self-serving to me. Slavery is one of the last things that comes to anyone's mind when thinking or talking about Napoleon. Napoleon's greatness does not stand on whether he sent or did not send slaves anywhere. To judge him on that count is like judging a doctor as if he were a pastry cook. And how is it that the BLM, or those who subscribe to it, get to measure Western leaders and heroes, culture, achievements, and marvels by the yardstick they choose? Did *I* agree to that? Did you? Has the BLM pulled the log out of its own eye yet? Now, the BLM is the very *last thing* I expected to see in an article about Napoleon. I get to see that movement's catchword on every 2nd person's lawn in the city where I live. And here it is again, holding up its sign in an article on Napoleon of all things. It's anachronistic, to say the least. I find most of, if not all, the anti-Western civilisation arguments these days, by whites or people of color, to be uncivil and incivil. As well as ungenerous, ingenuous, self-serving and dangerous, or some combination thereof.

I don't know anything about the Socialist

mayor Mayer of Rouen, he's probably a fine fellow, but it's a mistake for him, or anyone else, to be touching Napoleon's statue. Any statue of Napoleon. Anywhere. But especially in France, and especially today. Because after we in the West, by our own hands, have removed all the statues of our great men, and stopped reading, or started burning, their books, – because that is what is next! – and after we have effaced, eradicated, and expunged all hints of the greatness of Western culture and achievements off the face of the earth, – just to satisfy one group or groups of people who have a crowbar to wield and a chip on their shoulder, – at the expense of another group who likes their things, their culture, and their achievements – what will that leave us? It's clear: somebody else's stuff. Or maybe nothing at all. I'd like to challenge the BLM and M. Mayer-Rossignol to erect something new, better, brighter, original, – but without destroying something else in its place. Many of our Western lights were great, smart, good and talented men or women. We have *all* benefited from their accomplishments. It's plainly wrong to benefit from them while in the same breath disparaging them, to stand on the shoulders of giants all the while defecating on their pedestals, like so many Stymphalian birds rewriting history or trying to.

Rather than destroying the past, today's empowered masses, today's *Jacquerie,* should instead cherish the day and create their own new thing – statue, book, monument, whatever, – but in some *other* place, not some other's place. We can all coexist on earth, quietly and in peace. No one needs their temper raised, which often begets violence. When it's a question of a statue or a monument, a book or a billboard, that one does not like or approve of, we should just look the other way, turn the other cheek, like I do when I see the signs and slogans on so many lawns, in so many windows, *ad nauseum,* in my neighborhood and city, proselytizing to me, about something I don't subscribe to, and co-opting my nation. "In Our America..." *Our*? That includes me, right? Or did I give my neighbors and fellow citizens the right to speak exclusively on my behalf all of a sudden? No, I did not, but I leave the signs in place on their lawns and in their windows all the same. Or do the BLM and other denigrators of Western civilisation advocate throwing property and the freedom to think for oneself out the window, with the rest of the democratic bath water that is starting to stink? Let people have statues! Let them have their Rushmores too, and their *Arcs de Triomphe*. Let them have their history or histories. But let's stop the destruction of other peoples' culture and history, and the hypocrisy and self-serving claptrap that often accompanies it.[1]

I could say more, but I will say less. I may

---

[1] hypocrisy and self-serving claptrap: Witness whereof the news item that broke just days after this foreword was first written, reporting that BLM co-founder, Patrisse Cullors, had recently bought a $1,400,000 home in an affluent "white" neighborhood of Topanga Canyon (the Hollywood Hills basically) in Los Angeles, California, where "the vast majority of residents are white, according to reports." One has to wonder whether Ms. Cullors, originally from Pacoima, and a self-professed "trained Marxist," is still in favor of defunding the police now that she has purchased her new homes (yes homes, she owns more than four in the U.S. and several outside the U.S.); the BLM movement **is** however apparently still for defunding the police though, as their Webpage indicates; see https://blacklivesmatter.com/defundthepolice/. We hope that it works out for her and her many properties.

have said too much – but not as much as I would like to say; besides, I'd hate to sound whiny and ungenerous *like the people I softly contemn*; and my indignancy and rage for the daily assault on Western culture and achievements today – by Westerners and non Westerners alike, by whites as well as people of color – is probably not suitable for the foreword to a book.

In honor of the very great general, and in time for the 200th anniversary of his death, on May 5, 2021... *Vive l'Empereur!*

– Richard Robinson, April 9, 2021

The BLM "took in more than $90,000,000 last year" [2020]. That's a lot of money for an organization that "does not file financial disclosures" according to an article on buzzfeednews.com (July 6, 2020). Also, a $1.4 million home and "four high-end houses for $3.2 million in the US alone, according to property records" according to a *New York Post* article (April 10, 2021) is a lot of capitalism and private property for someone who "spent years absorbing Marxist-Leninist ideology."

When news of the co-founder's purchase of the $1,400,000 home was "twitted" (or is that "tweeted"?) on Twitter, the journalist who emitted the twit was locked out of his account on the social media platform. The BLM is said to have gotten its start on Twitter with its hashtag #blacklivesmatter.

The head of the BLM movement in NYC, Hawk Newsome, called for "an independent investigation" into Ms. Cullors and how her "global network" spends its money. He is quoted as saying, "We need black firms and black accountants to go in and find out where the money is going." (*New York Post* article, April 10, 2021). That is good, and so they should: we are all in favor of the investigation, and applaud Mr. Newsome for suggesting it, but one has to wonder why he feels the need specifically for *black* accountants and firms to investigate her. What difference does the skin color of the investigators make? *Iustitia cæca est*, no? Apparently not.

_____

The BLM movement, as indicated on the About page of the BLM Website (https://blacklivesmatter.com/about/), is a self-professed "global organization in the US, UK, and Canada [!!!], whose mission is to eradicate white supremacy..." The US, UK and Canada is hardly "global," but okay. It goes on to say that they "are a collective of liberators who believe in an inclusive... movement" and that they "affirm the lives of Black queer and trans folks, disabled folks, undocumented folks, folks with records, women, and all Black lives..." There is no mention of men, and no mention of Asian or white folk. That does not strike one as too very inclusive. And one has to feel somewhat ill at ease for all the women on earth, black or not, who get thrown into a group of otherwise undocumented immigrants, disabled people, convicts or ex-convicts, Black queer and trans folks.

It is very clear then that the BLM, _unlike_ Martin Luther King, for instance, or Mahatma Ghandi, is offering nothing new in terms of revolution and change. It's merely yet another push for regime change that peoples all over the world have seen time and again for the last 200+ years, where their leaders line their pockets, raise their walls, and do their utmost to grab and hold on to money, power, and influence. Their genius is to play on the guilt chord of young and not so young idealists, not of color, willing to cut their own throats and those of their children for a presumed level playing field. Would it be inappropriate to say, as Jesus did, "they know not what they do?" The BLM _do_ know what they do, however, and they have taken a page right out of the handbook of those they so contemn; they use it deftly with modern technology against the people they encourage to rally for them. That's about as inclusive as they get as far as Western civilization, men, and the Caucs go. And as they consolidate their power (they already have their money), they will do the same things, and make the same mistakes, as others before them have. It will be "The King is dead; Long live the King!" all over again. And Karl Heinrich Marx and Vladimir Ilyich Ulyanov Lenin, or what is left of them, will turn, sadly, once more, in their graves.

The Soul of Napoleon

"The whole world is the garment of my misery." – H. G. Wells. *When the Sleeper Wakes*.

Fortioribus fortior instat cruciatio.[2] – *Book of Wisdom*.

Dedication

TO ANDRÉ MARTINEAU

My dear André, it is not I who gives you this book, the most important, perhaps, of all those I have written to this day.

It is my son André who gives it to you, my sorrowful son André whom God took from me in his baptismal innocence and who is eighteen years old today, in Paradise.

He would have been the person I dedicated it to, and it is proper that you take his place in this manner. I want to believe that it is his wish.

He would have loved Napoleon as you love him, and your common Patron, the great Apostle of the Cross, will make you understand, if you interrogate him lovingly, what there was that was desirable and magnificent in the suffering of the most glorious of all mortals.

---

[2]*Fortioribus... cruciatio*: Latin for "But a greater torture awaits the very great." Book of Wisdom 6:9.

We are at the evening of the world, my dear child; you will be witness, perhaps, to divine and terrible things that the vanquisher of kings seems to have so grandiosely prefigured.

May the *The Soul of Napoleon* open your heart and serve as a comfort to you in unknown trials.

– *LÉON BLOY*

May 5, 1912

Introduction

I

The history of Napoleon is quite certainly the most unknown of all histories. Books that claim to recount it are innumerable, and there is no end to documents of every sort. In reality, Napoleon is perhaps less known to us than Alexander[3] and Sennacherib.[4] The more one studies, the more one discovers that he is the man whom nothing resembles and that's all there is. It's the unfathomable gulf. One knows the dates, one knows the deeds, victories or disasters, one knows a bit or quite a bit of the famous negotiations that are, today, merely dust. His name alone remains, his prodigious NAME, and when it is pronounced by the poorest of all children, it is enough to make a great man blush, no matter whom. Napoleon is the Face of God in darkness.

It is well-known that prophesies or Biblical prefigurations cannot be understood except after their *entire* fulfillment, that is to say, when all that is hidden will be revealed, just as Jesus announced in his Gospel, and that necessarily leads one to think above and beyond time. Napoleon is inexplicable and, without a shadow of a doubt, the most inexplicable of men because he is, before everything and above all, the Prefigurator of HIM who must come and who is per-

---

[3]Alexander: Alexander the Great, Alexander III, King of Macedon, 356-323 BC.

[4]Sennacherib: King of Assyria, 705-681 BC. He invaded Judah and is mentioned in the Old Testament.

haps no longer very far away, a prefigurator and a precursor entirely near to us, *signified,* himself, by all the extraordinary men who preceded him throughout time.

If one accepts this postulate and penetrates it a bit, one will see History take on a completely new aspect and the Napoleonic Ocean, so terribly turbulent until now, becomes all of a sudden very calm, under a sky of miraculous serenity.

Who among us, French or even foreigner, at the end of the XIXth century, – who has not felt an enormous sadness for the denouement of the incomparable Epic? For those with as little as an atom of soul, it was overwhelming to think on the fall, really too sudden, of the Great Empire and its Leader; to recall that one had been, only yesterday it seems, at the highest pinnacle of the Alps of Humanity; that by the single act of a Prodigious, Well-Loved, Redoubtable man, as has never been seen before, one could believe oneself like the first Couple in Paradise, absolute masters of what God placed under heaven; and that, so soon afterwards, it was necessary to fall back down into the old mud of the Bourbons! Truth be told, that fall had nearly uprooted the earth. The convulsions of 1813, despite the excessive grief and bitterness, were of such grandiosity that even imagination and pride can be consoled by them; but the end is too horrible, too sudden above all, once again, and the most angelic resignation is tempted to shy away from the doxology of that colossal Psalm of Penitence.

II

As we well know, immense mistakes were made, but those mistakes are precisely what make the sadness unbearable. What man, reading the history of the Empire, supposing himself a contemporary, has not tried to be persuaded for example that Napoleon would have less confidence in Russian loyalty, less caresses for Alexander at Tilsit; that he would demolish Prussia from top to bottom and re-establish Poland; that he would find a better expedient than the dangerous evasion of Bayonne; that he would not make kings of his miserable brothers; and not spread his forces so thin, from Cadiz to Moscow, squandering, destroying thus, the most beautiful armies in the world? To whom has it not occurred finally, in any case, to hope for the appearance of Grouchy at Waterloo, that mediocre and disastrous Grouchy so blindly chosen by the Emperor for the most decisive strategic action? And that's not all. How not to weep at the recitation of that second Abdication? The greatest of conquerors abdicating *two* times! Napoleon thrown down from his throne by a Fouché, by a Lafayette, then going to hand himself, body and soul, over to England!...

I stopped suffering for these things the day I was able to understand, or able at least to catch a glimpse of the completely symbolic destiny of that extraordinary Being.

In reality, every man is symbolic, and it is the extent to which he is a symbol that he is a living being. Certainly, that extent is unknown, as unknown and unknowable as the fabric of infinite combinations of universal Solidarity. He who knows exactly, by a marvel of infusion, what weighs down on an individual, that man there would have the entire divine Order like a planisphere laid out before his eyes.

What the Church calls the Communion of saints is an article of faith and cannot be anything other than that. One must believe in it as one believes in the economy of insects, in the exhalation of germs, in the Milky Way, knowing quite well that one cannot understand. When one refuses to believe, one is a sot or a degenerate. The Lord's Prayer teaches us that we must ask for *our* bread and not *my* bread. For all the earth and for all time. Identity of the bread of Caesar and of the bread of the slave. World-wide identity of impetration. Mysterious equilibrium of strength and weakness, hanging in the Balance where all is weighed. There is not a single human being capable of saying what he is, with certitude. Nobody knows what he has come to do in this world, to what his actions, his feelings, his thoughts correspond; who are his closest fellows among men, nor what his true *name* is, his imperishable Name in the register of Light. Emperor or stevedore, nobody knows his burden or his crown.

History is like an immense liturgical Text in which the jots and iotas are worth as much as its verses or entire chapters, but the importance of one or the other is indeterminable and profoundly hidden. If I think then that Napoleon could very well be a glimmering iota of glory, I am compelled to tell myself, at the same time, that the Battle of Friedland, for example, could have been as easily won by a small girl of three or by a centenarian vagabond asking God that his Will be done on earth as it is in heaven. So, what one calls Genius would be simply that divine Will incarnate, if I dare say so, having become visible and tangible in a human instrument brought to its highest degree of force and precision, but incapable, like a compass, of exceeding its extreme circumference.

That leaves, for Napoleon and for the entire infinite multitude of his inferiors, this: that we are all of us figures of the Invisible, and no one can move a finger nor massacre two million men without signifying something that will be manifest only in the beatific Vision. Throughout eternity, God knows that at a certain moment in time, known only by Himself, such or such a man will accomplish *freely* a *necessary* act. Incomprehensible accord of Free Will and Prescience. The most luminous intelligences have never been able to go beyond that limit. In such a state, integral Man, being unable, according to the creative Word, to be anything more than a resemblance or an image, renewable with each generation by a billion souls, is forced then to be suchlike forever, no matter what he does, and to prepare in this way, little by little, in the crepuscule of History, for an unimaginable advent.

There are clearly good men and bad men, and the Cross of the Redeemer is always there; but they both do only what is expected of them and cannot do anything else, neither coming into the world nor subsisting, that might alter the mysterious Text, while multiplying to infinity the symbolic figures and characters. Napoleon is the most visible of those indecipherable characters, the loftiest of those figures, and it is for that reason that he so astonished the world.

III

To tell the truth, the world is not difficult to astonish. It is so mediocre and low, that apanage of Satan, that a semblance of strength or grandeur suffices ordinarily. One has seen a lot of that in our own days when politicians and writers, capable at best of jabbing at cattle or plates, have been able to make themselves admired by the multitude.

Napoleon, gifted with strength and grandeur, more than any other man before him, must have astonished himself even more than all those whom he amazed. Aboriginal of an unknown spiritual region, a stranger by birth, and in career, in whatever country he found himself in, he was really astonished his entire life, like Gulliver in Lilliput, by his contemporaries' excessive inferiority, and his last words heard on the island of Saint Helena prove that that astonishment, turned into perfect contempt, was borne by him to the grave and before the tribunal of his Judge.

What then did he come to do in that France of the XVIIIth century that certainly didn't foresee him and expected him even less? Nothing more than this: *An Act of God by the Franks*, so that men all the world over would not forget that there really is a God and that he must come like a thief, at an hour one does not know, in the company of a definitive Astonishment that will procure the exinanition of the universe. It was appropriate doubtless that that act should be fulfilled by a man who barely believed in God and who was unfamiliar with his Commandments. Not having the investiture of a Patriarch nor of a Prophet, it was important that he was unconscious of his Mission, as much as a tempest or an earthquake is, to the point of being able to seem like an Antichrist or a demon to his enemies. It was necessary above all and before everything that, by him, the French Revolution should be consummated, the irreparable ruin of the Ancient world. Evidently God no longer wanted the ancient world. He wanted new things, and a Napoleon was necessary to inaugurate them. An Exodus that cost a million men their lives.

I have studied this history a great deal. I have studied it while praying, while weeping for joy or for sorrow, quite often asking myself, how many times! if it was not crazy to be reading it from a human point of view, as one reads the history of Cromwell or Frederick the Great, the only leaders, I think, who could be supposed after Hannibal or since Caesar to reside in the general vicinity of Napoleon, and I ended up by feeling that I was in the presence of one of the most redoubtable mysteries in History.

IV

A young man arrives on the scene who does not even know himself and who must think himself infinitely far from a supernatural mission, – if even the idea of such a mission can occur to him. He has a sense of war and ambitions a military situation for himself. After many miseries and humiliations, he is given a poor army unit and, suddenly, shows himself to be the boldest, most infallible of captains. The miracle begins and never stops.

Europe which had never seen the likes of him begins to tremble. That soldier becomes the Master. He becomes the Emperor of the French, the Emperor of the West, – the EMPEROR, simply and absolutely for the duration of centuries. He is obeyed by six hundred thousand warriors that no one can vanquish and who adore him. He does what he wants to do, remakes as he pleases the face of the earth. At Erfurt, at Dresden above all, he seemed like a God. The potentates licked his feet. He attained the sun of Louis XIV, he married the loftiest girl in the world; supercilious and wrinkled Germany hasn't got enough bells, cannons, or fanfares to honor that Xerces who remembers with pride his having been sub-lieutenant of artillery twenty-five years earlier, never to have possessed a *sou* nor *stitch* and who now leads twenty peoples in the conquest of the Orient.

A season passes and here is "the cold North Wind that devours the mountains, *sicut igne*,"[5] says Ecclesiasticus. The sub-lieutenant of 1785 returns on foot through the snow, supported by a stick, followed by several dying men. But he is vanquished only by heaven, not yet ready to be vanquished by men. God loves that proud man and afflicts him with love, without wishing to demoralize him completely. God had looked into the liquid blood of carnage and that mir-

---

[5] *The cold... sicut igne*: A loose translation in the original, from Latin, (except for the "sicut igne" which is in the original Vulgate) from *Ecclesiasticus* 43:23.

ror reflected back at him the face of Napoleon. He loves it like he loves his own image: he cherishes that Violent man like he cherishes his most beloved Apostles, his Martyrs, his Confessors; he tenderly caresses him with powerful hands, like an imperious master caressing a timid virgin who would refuse to get undressed before him. He will strip him down in the end certainly and in so complete a manner that kings will be busy, for thirty or forty years after the fact, arguing over his rags. But he does not want that at first. He will try again, three times even. 1813, 1814, 1815, three Epiphanies of suffering!

The first, and not the least terrible, is that which most resembles the deluge of the Vth century. The colossal armies of the supreme Coalition repeat rather well the Huns, the Sarmatians, the Suebians, the Alans, the Saxons, the Goths, and the Vandals in their Punishment of Rome. The entire pack of barbarous dogs is at the flanks of the mutilated, but undefeated Lion. He withdraws, roaring with pain and pride, and returns to France where he makes them fight, one against ten, the children transformed by him into legionnaires. Mount Olympus or the Valhalla of imbecilic Gods trembles once again. Betrayed finally by lieutenants whom he had conceived and raised, he is relegated to the derisory island of Sancho Panza. All seemed finished. It all seemed over. A fratricidal and libertine old man tried to gnaw France with his gums. The Invincible One re-appears one last time, how prodigious!

The Kingdom of Jesus Christ and his Mother, depleted of blood, crippled by sadness, rushes imme-

diately towards him letting out cries of joy. It's 1815,
alas! and Waterloo! They fight like angels in despair.
They fight against all History, against sixty centuries!
It's a disaster, and Joan of Arc weeps on all the path-
ways. Napoleon, who brought victory with him, is
forced to hide it in the bushes during the rout, not
wishing to be vanquished by anyone but himself. In-
comprehensibly, he abdicates a second time, disgusted by everyone and everything, and ends up on Saint
Helena in the midst of England's rats and scorpions.

V

Such is that historical mystery, unlike any other. Formerly, in the time of my youth, and even later, when I
loved reading adventure novels or melodramas, I no-
ticed that what excited me most was the *incertitude
over the identity of people*. It's a great resource, inex-
haustible even today, in pathetic Fiction. Since Œdi-
pus and Jocasta, that has not changed. It is essential
that the hero, somewhat more intuitive, moreover,
than one wants to imagine him, should be an enigmatic individual himself. That power of a banal idea,
which cannot be shaken, clearly draws on some deep
presentiment. It's the effect of a direct, but very ancient, view into the human condition. I have said it already, – each man on earth is meant to signify something he is ignorant of and to realize in this way a par-
cel or a mountain of invisible materials by which the
City of God will be built. Not to see in Napoleon anything more than a man greater than others, assuredly,
but signifying nothing beyond his acts, is to invalidate
both the Future and the Past, disqualifying thereby all

History.

"*Ego dixi, dii estis.* I said: You are gods," said the Master. Ah! without a doubt, men are, at the very least, figures of God, monstrances of his mystery, and Napoleon is certainly the most manifest there is that it is possible to contemplate. I do not believe that there was, in all his life, an action or a circumstance that cannot be interpreted divinely, that is to say in the sense of a prefiguration of the Reign of God on earth.

He is born on an island. He constantly makes war on an island.[6] When he falls for the first time, it's on an island. Finally, he dies captive on an island. Insular by birth, insular by emulation, insular by the need to live, insular by the need to die. Even when he held Europe in his grip, even in his worst battles, the perpetual rumbling of the Ocean would drown out the sound of the cannons for him. Ambitious to reign over all the seas, the continent was always an obstacle for him.

Like a large marine vessel caught in the ice, he was continually caught on land and could not succeed at disengaging himself. For twenty years he trudged over the continent with a fury, not forgiving it for getting in the way of his conquest of that inaccessible English isle, over which he would have been the sure master of the Atlantic and of the Mediterranean, shutting in with his fleets the old realms and old empires and making an island of all the earth, another island, as big as his dream! *Tacete et ululate, qui habi-*

---

[6]makes war on an island: scil., England.

tatis in insula,[7] he seemed to say along with the Prophet, with each step he took, and it was in vain.

VI

He decrees the continental Blockade, the most enormous ever conceived or undertaken. The entire European continent reclused and locked up, three hundred million men, if necessary, condemned to ruin and despair, so that England, banished from the people, should be forced to hand over the keys and unfetter the triple-barred jail of the oceans, and it didn't take long... That recalls, on a very grand scale, the famous interdictions of the Middle Ages, memory of which is so troubling. Apocalyptic decree! One imagines it dated from the day before the universal Judgment. There are angels and clarions in every canton of the sky.

But the Scythians and the Sarmatians[8] are just beginning to wake up to Western civilization. Is it not right that they should have the time to putrefy in their turn? They refuse to be immolated. Napoleon falls on them with ten armies. Now watch how God protects his barbarians. The fabulous and invincible warriors are killed by the cold, and the Blockade becomes impossible. Impossible also, from then on, world Domination.

---

[7] *Tacete... insula*: Latin for "Be quiet and howl, you who inhabit an island," which is a combinatorial quote from Isaiah, 23:2 and 23:6.

[8] Scythians and Sarmatians: in Southern Russia, and Southeastern Ukraine.

It was beautiful however, too beautiful doubtless for that jealous God who does not want to share. When he deigns to manifest himself at the end of ends, that is to say when all figures will have been exhausted, it will be quite necessary that he do something similar to Napoleon's Design. Then, and only then, will one know just how beautiful it was! Certainly, at that moment then, God will have before him and against him an island to humble, to exterminate, the *Island of Saints*, formerly, become since then the tragic and somber island, the island of Renunciation, Apostasies, Hypocrisies, Betrayals, and Pride. It will be quite necessary then that, in a certain way, he separate it from the continent of Faith, itself already sequestered in perfect exhaustion and stupefaction.

Because it will be necessary, O Jesus, you who call yourself, in our own words, the Son of Man, that you should be contented with so little, if, in the meantime, you have not miraculously changed everything. Given it is inevitable that all figures should be fulfilled, you will have, just as your Napoleon did, the obstacle of cold weather and barbarians. But, at the same time, you will have the resources he didn't have, to turn his people into something, like a new people who will be ranked just below the angels.

VII

Napoleon marries twice, like an Ahasuerus,[9] repudiating a prostitute in order to take up with another wom-

---

[9] Ahasuerus: the name of a Persian (or Babylonian) king, potentially a reference to Xerxes I, 485-464 B.C.

an who had nothing in common with Esther in the Bible, except the perfumes. But she was from among those of the Caesarian monarchy of Hapsburg, old evanescent bergamot that appeared to intoxicate him one day, on account of which he was soon stunned and staggering, almost asphyxiated, dangerous effluence of ancient sepulchers of carnal magnificence and grandeur.

It is said that Ahasuerus who reigned over twenty-seven Asian provinces, wishing to replace his first wife, had the most beautiful girls in the world sought throughout his empire and compared, "those even of Parthia and indomitable Scythia," and that in the end he fixed his choice on a poor Jewish girl named Esther, which means *Mysterious*. Napoleon, more powerful than that ancient potentate and not wanting a poor woman, had to choose among the loftiest heirs of the Majesties who licked his boots, and he went about it as if leading a rapid campaign, sweeping away the princesses of lesser grandeur with a wave of his hand. But she whom he married was certainly not a mysterious woman, and the vile father-in-law, the man with the "viscera of State," as his domestics said, having become, four years later, a Mordecai of adultery, conducted, in person, with the three crowns on his head, his archduchess of a daughter to the brothel, in order to dishonor his son-in-law who did not make him tremble anymore.

As a parting comment, and not leaving the Bible, one would think one was reading Ezekiel in that formidable chapter where the nameless ignominy of the two spouses of the Lord is divulged.

VIII

Shall we speak of the return from Elba Island. What has not been said or written about that incomprehensible event? Up until then, Napoleon had conquered only men, and precisely because he was greater than all of them, he had been or appeared to have been vanquished in the end. But on quitting Elba Island, he undertakes to combat the nature of things, his own destiny, forcing himself to overwhelm the formidable Angel, like powerful Israel against God even. No one had seen anything, and no one will ever see anything perhaps, comparable to the flight of the eagle going from "steeple to steeple to the towers of Notre-Dame." Why Notre-Dame? Napoleon was not devoted to the Holy Virgin, not ostensibly at least. But with everything being presumable with respect to a being who was so great, is it not permitted to suppose in him a superhuman presentiment, a secret divination, of the Suzerainty of Mary, Patroness and Protectress forever of that France that he had picked up out of a mud hole of blood and filth and which he had made so magnificent?

And now, watch how I astonish myself with my prudence! Why so many literary precautions? Does that not blind your eyes that the Event was completely and absolutely supernatural? There was not a family in France perhaps that did not bleed to the point of fainting, to the point of a definitive stopping of the heart. In Italy, in Egypt, in Germany, in Poland, in Spain above all, and in Russia, an infinite number of French had died voluntarily or one could believe voluntarily. The Saxony campaign in itself cost more than one hundred thousand lives. One would have thought that that unappeasable devourer had extenuated every enthusiasm and dried up every fountain of love.

It was just the opposite. A last army of victims came forward and offered itself up, and what victims! A roaring of glory mounted to the gates of heaven. During a review, the heroic mounted soldiers of one hundred battalions, crossing their sabers above his head, made a vault of steel for him while weeping for joy and uproariously. Several days later, they were immolated in turn. They were the last, but it didn't make any difference, and Napoleon, if he had wanted to, could have still, even after Waterloo, continued the human sacrifices indefinitely.

In truth, a man has never been so worshiped as that man there, in hope or in despair, in the immense torments of fatigue, in hunger and in thirst, amidst the mud and amidst the snows, in the hail of bullets and the licking of flames, in the exiles, the prisons, the hospitals, and among the agonies; worshiped all the same, worshiped ever and always, and forever on end, in spite of everything, like a redeemer whom the corruption of the tomb could not touch, like a virgin of glory who cannot die. I have known, in my childhood, old disabled veterans who were incapable of telling the difference between him and the Son of God.

IX

What an impression those images of Raffat illustrating the poor story of the Norvins which appeared to me like a gospel when I was twelve years old! A gospel, – that was it exactly. I hardly knew any other, my Christian cultivation having been outstripped, or retarded, by the Napoleonic cultivation. Despite many years having passed, I still find a magnificent shiver that runs through me when flipping through those pages which I was barely able to read then, completely ignorant of the history. But what fever, what trembling because of those images! What did I need to read? With them and through them, I followed him everywhere, my hero and my emperor, from Toulon to Saint Helena. I accompanied him above all into Egypt and into Russia: I saw him always all-powerful, always infallible like a God, and I believed myself to be one of the oldest of his Old Guard.

What did I need to understand? I already felt, and I never stopped feeling in him, the Supernatural, and the eight letters of his name, imprinted, I remember, in large capital letters the color of blood, on the cover, – they seemed to me to emit rays of light to the ends of the universe. I have never gotten over it.

There was also, very close to the town where I lived, a strange garden and certainly very ridiculous that I will see again in Paradise maybe. A bourgeois whoever, an imbecile, I'm afraid, had imagined to turn his property into a place of Napoleonic pilgrimage. It was called Saint Helena and my father brought me there as a child. It's so long ago that I can barely remember it. There was, I don't know what, an enormous bust of the Emperor, a small column of the Grand Army in imitation bronze, a sort of cavern sur-

rounded by weeping willows and representing the
tomb in exile, from which emanated a religious terror,
rocks from Malmaison or Saint-Cloud, a greenish ef-
figy of the King of Rome in a cradle of ivy or honey-
suckle, and plaster casts of grognards or marshals de-
fying every sort of comicalness under the moon.

That's all I can find again in the crypts of my
memory and still I'm not very sure of it. But the emotion in my heart of a child lasts forever and it is because it has not stopped, for fifty years now, that I can
write these pages. Such was and such is still, so long
after his last sigh, the powerful influence of that
Prodigious One!

X

To envisage Napoleon as a divine instrument puts one
at ease in order to speak of his faults, recorded with
so much effort and on so much paper by all his
judges. If one reasonably understands the word faults
to mean a series of voluntary transgressions, venial or
capital, of a promulgated law, strict justice does not
allow one to impute them to an instrument. In this
sense, Napoleon cannot have committed a single
fault, having always been forced to fulfill, in his capacity as an instrument, what was prescribed to him
to want or to accomplish.

There is no doubt that he was, at the same
time, a man under the law of the fall, accountable, by
consequence, for the mischief of his freedom. But of
that, God alone is the judge, *Dios de todos*. I have

only seen what one commonly calls political faults. Nobody else but him can know or conjecture without being reckless what he put of his own will into the magnificent or dreadful actions required by a superior Will which he had to obey.

Confusedly, he felt it when he spoke about his "star." Unable to understand, he felt a Hand on his head, a hand on his heart that stopped beating then, it was said, a hand around his formidable thought. Shuddering, that Master of the world saw himself circumscribed in a freedom of inferior order and – beneath his imperial mask – the *cadet*, in that way, compared to all others, the most miserable even, who did not have, like him, orders, a mandate of eternity, a divine canvas to fill, and who appeared to have, more than he did, a choice in their good or evil works.

Perhaps it would be possible to explain then, by the intermittent rebellions in his soul, by his sudden whims to escape so fatal a grandeur, the strange forgiveness he had shown so many times to his most dangerous enemies and his inconceivable weakness for companions unworthy of him.

"That man born for empire," wrote a historian of penetrating insight, "who entered easily into his sovereignty and effortlessly found himself not only the equal but the superior, and in all aspects, to the kings and emperors vanquished by him, – he always remained an upstart and a cadet to his family. In that respect, he was never emperor except when giving. He never succeeded in having himself obeyed nor respected. He preserved, for his family, that strange complaisance that he extended to all those who had assisted him in difficult times, served him in his years of crisis. That warrior, that violent autocrat, generous, debonair, was of all masters and leaders of men, the most notably fooled and betrayed by his wives, by his brothers, by his sisters, by his ministers, by his lieutenants, by his servants."

It is clear then that he needed to be like that, and even his worst faults, given one must employ that word, were like the essential parts of the poem of his destiny.

XI

One is, anyways, sufficiently warned when, being capable of profundity, one comes to consider the palpable sottishness of an imaginary substitution for accomplished events. "An altogether different denouement would have taken place," it is said, "if such a circumstance had been foreseen." But precisely that circumstance could not have been foreseen nor brushed aside because that denouement, and no other, was necessary. The facts are absolute in themselves and in all their peripeteia. The historical facts are the Style of the Word of God and that word cannot be conditional. Vincennes was necessary, Tilsit and Bayonne were necessary, the brother Kings were necessary, the incomprehensible impunity for Bernadotte and the disastrous campaign against Moscow were necessary; after Dresden and Kulm, the incommensurable madness of abandoning, in the useless German fortresses, the more than 150,000 soldiers, more than enough to crush the Coalition on the plains of Champagne, was necessary. Grouchy, finally, was necessary. All these things, and many others, that one does not know, were necessary, and the proof without rejoinder is that they happened in plain sight of God who does not make errors and who wanted these things from time immemorial.

"Have I fulfilled then the wishes of Destiny?" responded the Emperor to some of his great men who attempted, in 1812, to dissuade him from Russia. "I feel myself *pushed* towards a goal that I do not know. When I have attained it, an atom will suffice to put me down." He defended himself, at Saint Helena, of the reproach that he loved war too much, saying that he had always been constrained to it, and that is absolutely on the mark. If he loved the war that he excelled at, where does that put the great artist in love with his art but forced to live by it exclusively, – who would have the right to incriminate him?

One wonders what man has been able to ride under the whip of his destiny as far as he did. Everyone is familiar with his famous ride at top speed, from Valladolid to Burges, thirty-five leagues in five hours. He had left with a numerous escort, because of the danger of guerillas. With each step of the way, he left the world behind him and arrived almost alone. He had to leave the English, insufficiently throttled in the north of Spain, to throw himself on the menacing Austria, and there was no time to lose. That fantastic, almost unbelievable, ride is a perfect image of the entire, frenzied life of that Titan, always constrained to go one step in advance of the lightning, and who did not obtain rest except in death.

XII

For attention deficiency or debility of intelligence, I am often surprised by the two Abdications, unable to conceive how such a man would have abdicated once even. I think, today, that he did that like everything else, by commandment. It's another version of the two spouses. I'm at the point of telling myself that that there is above all where one must look.

Would it be possible then to have two divine abdications? Is such a thought even conceivable? God saying: "From now on, I am no longer God." A first time, because he is abandoned, a second time, because he has abandoned himself. It's vertigo, it's the cliff of the absurd and impossibility. And be that as it may, it happened, in the great mirror of enigmas, in 1814 and 1815. So many people wept over it and there are people who weep still. Before and primarily after the Hundred Days, the poor souls said to themselves: "It's over! We have no more God, what will become of us? No one will be born, no one will die. No one will be judged nor recompensed by anyone. No more paradise to hope for, no more hell to despair of." And there was in the world of the poor a boundless sadness.

Why then did Napoleon abdicate and, I repeat, abdicate twice? Only the Holy Spirit could answer that question. "It's for me," it would say, "that he abdicated. Bearing a resemblance to the Father who has repented of having made men, being in the image of the Son crucified by them, Napoleon was forced to dismiss them in person and in that way, given there was nothing more to prefigure than the Paraclete of definitive triumph in which all symbols must be fulfilled and all prophesies consumed. Your emperor did what he had to do, to the letter, like suns or animals, without understanding or knowing, and the magnificence that appeared in him before he tumbled, was, by anticipation, merely an infinitely pale reflection of my coming splendor. The two gestures by which he has left you were *mine*, really, in space and time, but in a mode that is hidden to you and that you cannot know before the proper hour."

"*Let Him who can understand understand*,"[10] said Jesus, who spoke only in parables, and that mysterious injunction *could only* be addressed to the Paraclete that was to come, by which all the arcana will be disclosed.

Not being the accredited proxy of that Consoler, I have therefore nothing to explain. Besides, since the decline and abjection procured by the original Fall, who is capable of explaining or deeply understanding whatsoever it might be? It is already amazing and passably superhuman to show that there is mystery everywhere or to give a foreboding of it; to proclaim, for example, that there are no cases judged in history, that the life of men, great or small, "is not taken away, but only *changed*," according to the liturgical expression, *vita mutatur, non tollitur*, and that, in consequence, one knows nothing really of the perpetually iterative combinations of the divine Will!

---

[10]*Let... understand: Matthew* 19:11.

XIII

Ah! if Napoleon could have been the multitude! If his Name had been the name of the multitude, how much easier that would have been to explain him! To start with, he would not have been born on an island, which would have simplified everything, his case being essentially geographical, and every idea of a continental or only departmental Blockade would have been without occasion and without opportunity. A single spouse would have sufficed, universal Sottishness, faithful wife if ever there was one, and how very fecund! He would never have been on Elba Island, too far from the centers and, by consequence, he would never have had to return. As for the two Abdications, let's not even talk about it. It would have been easy to replace them by the universal Suffrage that would have certainly, we hope, terrified the Coalition and one would have had political prostitution fifty years earlier than we did.

But here's how it fell out. Napoleon was not the multitude. He was alone, absolutely and terribly alone, and his solitude was an aspect of eternity. The famous anchorites of Christian antiquity had, in their deserts, conversations with Angels. Those saints were isolated, but not *unique*; they saw each other sometimes, and their number is difficult to count. Napoleon, similar to a monster that had survived the abolition of his species, was really alone, without companions to understand him or assist him, without visible angels and, maybe also, without God; but who can know that?

Having no equals or similars, he was alone

amidst kings and other emperors who resembled domestics as soon as they drew near his person; he was
alone amidst the great men of his court whom he had
fabricated out of the mud and spit, and who returned
to their origin, the very day when his power began to
decline; he was alone amidst his poor soldiers who
could give him only their blood and who were not
stinting of it. He was alone on Saint Helena amidst
the Longwood rats and the gnawing devotions that
pretended to console him. He was alone finally and
above all with himself, wherever he erred like a lep-
rous untouchable in an immense and deserted palace!
Alone forever, like the Mountain or the Ocean!...

Chapter 1: The Soul of Napoleon

The first of all rights Napoleon could claim certainly, as well as the last drummer of his armies, was to have a soul, a soul that was really his own and that could not be shared with anyone else. It is difficult to think about.

Doubtless, when one is a Christian, one is required to know that all men have a soul, and that that invisible creature bears a resemblance to the invisible Creator. One knows also and by consequence that the soul of anyone, it does not matter whom, be it an imbecile or a negro, is infinitely more precious than all imaginable treasures, incomparably more colossal than the star Canopus which the most moderate astronomers concede has a spherical dimension eight million times greater than that of our sun. The saints have said that if someone could see a soul such as that, in its grandeur and in its dignity, he would die on the spot. Assuredly, if there were any doubt about it, the Dogma of Redemption by the Blood and Opprobrium of a incarnate God would be absurd and inconceivable.

It is already a great deal for a believer that the Soul could be thought of, and it is even completely supernatural, I dare say, that it should be spoken about continually. It is not a question here, of course, of the soul of animals or plants, which is their life principle, and which is really not easy to explain or

demonstrate. It is a question of the human soul incapable of ending, whose very existence is not known
except through an act of Grace, of the invisible soul
having to outlive a visible body that it is called upon
to reintegrate with one day, of that Soul which God
made a participant of Himself and which is more
durable than all worlds put together.

If that idea is overwhelming to us, when our
mind deigns to busy itself with the first person who
comes along, what might it be like for a Napoleon?
Will one need to say, while poking fun at the Redeemer and his Blood, that the soul of Napoleon is
more precious than that of others? Assuredly not, but
greater and incomparably greater by attribution, that
much is certain.

There are souls that are spouses or preferred
concubines whom the Lord is pleased to shower with
the most extraordinary and sumptuous of finery. If
they are disloyal or prodigal, they will incur punishment because the Master is as jealous as he is powerful. But even in the depths of their disgrace, they will
hold on to their essential glory and the memory of
what they were will not be erased from the heart of
men.

No one will blaze as brightly as Napoleon did,
that is for sure, but nothing proves that his soul was
more *illuminating* than that of a prig or a cobbler. The
lamps or beacons of his genius emitted a dazzling
light that lasts to this day and will not go out save on
the dawn of the Day of God. But his soul, ever a mystery, could shed light only on himself in a way that
we have no idea of. His very own soul, sad or joyful,

dark like the abysses, or tortured by the light; his soul of a sinner, of a proud, implacable, sentimental, and debonair man; his soul with its changing fires, sorrowful or triumphant; his inconstant or desperate soul, always saying to him: "You are alone, O Napoleon, eternally alone; nobody accompanies you, nobody knows what you love nor what you hate, nor where your steps will take you, because you yourself do not know. Poor, all-powerful and miserable wretch, weep under my wing, I will hide you and protect you."

Napoleon had nothing that was his own except his soul. It is because of it that he won all his battles; it is because of it that he was an extraordinary leader of men, a tireless administrator; that he dared to knead Europe in the hands borrowed from God and which he hoped never to have to give up again. It is through his soul finally and his soul *alone* that he had the glory to be mistaken like no man before him, and to be beaten in the end, being only the Herald, not through the furious hostility of several humiliated kings, but through the coalition of all the centuries and by the ebb tide of the French Revolution that receded before him, having borne him to the heights.

The historical testimonies are clear enough. Configurator and Regulator of that Revolution that changed the face of the world, Napoleon had against him, necessarily, all the anterior Traditions. All the things of the Past had to rush at him and over him, like innumerable torrents drawn to a single gulf.

Vainly he tried to harness them for his own purposes, displacing all the frontiers, trying to crown new kings and create new peoples, dating a new era from his person. Things obeyed him less than men did, and it boggles one's mind to tell oneself that he had a soul, a single soul containing pride, love, and suffering like others, to feel that, a soul excessively enormous, but absolutely one of a kind by destination, in which it was necessary to concentrate the effort of continual resistance to all other souls, perfidious mares or savage steeds, that it was indispensable always to dominate.

At the risk of sounding paradoxical, I dare mention the word disinterestedness. What could possibly be, in effect, the interest or the *interests* of a man having arrived at so prodigious a situation? What ambition could he have conceived, if not to be or to remain what he was already, what he was always meant to be, even in the limbo of his destiny, because the future, in the ordinary sense, is a word without acceptation, when one speaks of such paragons of humanity. At the acme of everything, from the moment he was thirty-eight years old, having had his fill of everything that could inspire a man, the only thing that remained was to make himself worshiped like a pagan king, if his inordinate power had been capable of overcoming the drop of water of his baptism.

The disinterestedness of Napoleon! But who thinks about that? It was within his capacity however, and completely out of proportion, not exactly through contempt or satiety, but because he didn't have the time to research or even to consider what could have been profitable for him. He had the disinterestedness of a true soldier who executes a dangerous assignment without being sustained only by the thought that his obedience could appear heroic. He himself not knowing where a mysterious Will carried him, which he didn't dream of disputing the exigencies of, and keeping only for himself the entire responsibility that a mortal could assume, it appeared simple to him to demand absolute disinterestedness in the many millions of creatures whom he showered with his glory, having nothing else to give them; but quite well divining that those inferior instruments of irresistible Force, which he was subject to the impetus of, went like him, and at the same pace, to the ineluctable fulfillment of a Design that surpassed the comprehension of his genius.

I can never say it often enough, – everything was against him, all souls against his one soul! Not only the souls of contemporaries so violently held in check by him, but the souls of the past, the souls, ever living, of dead ancestors who had filled drop by drop, throughout the centuries, the Seven Cups of Anger which he was charged with offering to the world, and even the souls to come, over whom those dreadful Cups would inevitably be poured, for he was nothing, as I have said, but a Precursor. Everything, once again, was bound to be against him, like criminals against the executor of their works, and also by virtue of that universal instinct of humanity in a state of decline which does not forgive its Superiors.

It is reasonable then to think that Napoleon, even on the days of his most dazzling triumphs, was a secretly but deeply unfortunate man, because happiness in this life, or what one wishes to call happiness, is nothing but a trick or a chance occurrence, illusory

in all other respects, of mediocre satisfactions and ad-
ventitious windfalls that cannot suit a great man and
above all not the greatest of men.

Chapter 2: The Other Souls

The number of them is infinite, and it is discouraging to think about them. The *other* souls, they are the human race altogether. For such is the dazzling effect procured, I do not say through revery, but through thought. Napoleon on the one side, the world on the other.

It seems to me as though I have lived through that period, not yet forgotten, from the year VI, when Bonaparte brought the ratification to Paris, so in vain! of the treaty of Campo-Formio.[11] It was the springtime of delirium, the beginning of universal fascination. Men would do anything to be seen next to that young general with the face of an ancient hero and comparable only to imaginary conquerors, he who had just, at twenty-eight years old, made kneel before him the classic armies of Austria which, not one hundred years earlier, were victorious over Louis XIV. One could almost not even breathe any longer amidst that great people exuding with glory.

From that moment forward, the dominator must have sensed his strength and assessed his contemporaries. Assuredly, he had to have seen how easy it was, with his gifts, to trample underfoot whatever there was that was greater than him, whatever one thought was greater throughout the centuries. Then,

---

[11]Treaty of Campo-Formio: a peace treaty signed, Oct 17, 1797, between the Republic of France and Austria.

necessarily, it must have begun for him, and already against him, the spectacle, until then unknown, of the furious avalanche of all souls inhabiting or having inhabited bodies for so long a time or since the beginning of time.

Without going as far back as the Deluge, there was, at least, Henri IV, the Gascon king, destroyer of Catholic unity in France and absurdly ambitious for a European hegemony that did not make allowances for the providential knife of Ravaillac.[12] That braggart of the "chicken in the pot," who was unable to leave the people with any other memory of himself than his ribaldry, he had dared to say, feeling himself threatened: "You do not know me, you others; when I am gone, you will recognize what I was worth and *the difference there is between me and other men.*" He believed it doubtless, and his grandson[13] believed it even more than he did.

The protocolist Louis XIV, supreme chief of the bureau of monarchies and one of the most mediocre swells ever seen in history, not judging himself the "unequal of many suns, *nec pluribus impar,*" insisted simply that one became blind or an idiot even when looking upon him. The Bourbon Louis XV, very worthy of his ascendance, just after his death, O Juvenal! had to be hastily put on bier with a cesspool emptier's pomp,[14] and that is the most characteristic trait of his reign. Finally Louis XVI, the Nothing,

---

[12]Ravaillac: François Ravaillac, the French Catholic who assassinated French King Henry IV in 1610.

[13]grandson: Louis XIV, the Sun King.

pneumatic and automatic royal, a killer of swallows
and a locksmith, capable, what's more, according to
Thiébault, of knocking out small dogs by blows of a
stick and laughing at that good farce, uncontrollably;
excellent object for the guillotine and inappreciable
treasure for diptychs on the martyrology of imbeciles.

It is understood that those personages, with all
their next of kin, friends, ministers, wives or mistresses, had souls. One is compelled to say the same about
each of the great mimes of the Revolution, from
Mirabeau all the way to the green-hued Robespierre.
And, when Napoleon had stopped barring the space
that existed under heaven, the story continues ignobly
with that sack of excrements called Louis XVIII and
his imbecilic younger brother Charles X, both fratri-
cides and disgusting supplanters of their nephew the
unfortunate Louis XVII, as little capable those last
two of a flash of superior intelligence as of an act of
courage or magnanimous kindness.

There would be no end to the prostitution of
the imagination if one had to speak about Louis-
Philippe, or the capitulator of Sedan,[15] or the Presi-
dents of our whore of a Republic, or above all of the
Monster whom one already hears banging on the window panes of the inn.

I have said that Napoleon is precisely at the

---

[14]bier... pomp: pomp or pump. Apparently Louis XV's body was
not embalmed and his heart was not mummified, according to
protocol, and his "pestiferous remains" were placed on the bier
as such, which caused a stink literally and figuratively.

[15]capitulator of Sedan: Napoleon III, in 1870, during the Franco-
Prussian War.

center of that immense whirlwind, unable to be anywhere else because of the exorbitant grandeur of his
soul. At that loftiness of thought that I am striving to
attain, it is clear that the notions of time or space no
longer exist. History as a whole becomes synoptic
and simultaneous, to the point where it is possible to
juxtapose and closely annex, under one's view, the
most disparate or distant events. Duration of time is
an illusion consecutive with the infirmity of fallen human nature. "Every man is the addition to his race,"
said a philosopher profoundly. Every great man is an
addition of souls.

In a distant and passably obscure epoch, there
was a moment when all that one calls the Past had a
need to come to an end with Charlemagne. Similarly,
one hundred years ago, it was necessary that everything, Charlemagne in the lead, rushed at Napoleon,
and that conflict is doubtless the most extraordinary
among prodigies. It is therefore inevitable to affirm
that Napoleon is the sovereign Leader of all wills, anterior, contemporary, or posterior and that he central-
izes in his own soul the totality of all souls.

In that sense and after an ideal defalcation of
chronological appearance, one can say – without
wearing a straitjacket – that Louis XIV, for example,
lacked deference with respect to Napoleon by making
a king of Spain out of his duke d'Anjou, after having
scandalously *disobeyed* him by signing the deplorable
treaty of Ryswick.[16] How many other things besides!

[16]treaty of Ryswick: a series of peace treaties signed in the Dutch
city of Ryswick, from September 20 to October 30, 1697,
between France and the "Great Alliance" of England, Spain,
Austria, and the Dutch.

The inertia of that miserable Christian sultan after Steinkerque, when he could have crushed Guillaume d'Orange; the savage and useless setting on fire of the Palatinate; the stupid expulsion of two or three hundred thousand Calvinists whom it would have been easy and so refreshing to massacre; the even stupider still the bombardment of Algiers and Tunisia not leading to conquest, and the pointless peace of Nimège, occasion for the bourgeois of Paris to rig out the triumphant victor in periwig with the nickname of "the Great," at the very hour when that political ploy, at the same time that it cut into France's prestige, prepared, at the end of the following century, for the future coalitions and England's definitive victory.

In sum total, Napoleon owed to him the discomfiture of Trafalgar, the anguish of Austerlitz, the mourning of Eylau, the illusion of Tilsit, the dishonorable treachery of Bayonne and the atrocious setback that resulted from it, the dreadful danger of Essling, the insane Marriage, the extermination of his power in Russia, the bottomless pit of 1813, the despair of 1814, and the final crushing defeat of Waterloo.

All that, and his mortal Captivity, is certainly owing to the ridiculous sun of Louis XIV, to the pale and obscene moon of Louis XV, to the confused pumpkin of Louis XVI, and finally to the enraged propulsion of the Committee of Public Health tending to exceed all bounds without possibility of retreat. Heir and testamentary executor of all those dirty or tragic souls, Napoleon had to go as far as Moscow to defend the barriers of Paris, and that was the catastrophe.

In his eyes, directly, who were the souls? Everyone thinks naturally of Talleyrand, Fouché, Bernadotte whom no opprobrium could condemn enough. But there were his bitches for wives, there were his brothers and sisters, all those whom he had made great, the immense pack of functionaries whom he had elevated in stature, the very nation become by him queen of the world. Then, in the crepuscular future, all that we know, alas!... Finally, one has to ask oneself whether it is really possible to conceive of a more torturing destiny!

Chapter 3: The Anguish

The most difficult moment in Napoleon's entire life appears to have been 18 Brumaire or, more exactly, the day after that celebrated coup d'état was consummated; after Bonaparte, horribly jostled by the Jacobins of the Council of Five Hundred and rescued from their hands by some of his grenadiers, had finally violated fortune by expulsing the Assembly.

Doubtless he had many other cruel moments and in greater number than one can imagine. But this started him down the path to emperor. For the first time, he needed to reach out and grasp the symbolic Globe, and he saw himself very near to perishing ignominiously. His ears were full of and buzzing with the terrible *Hors la loi*! Jacobin clamor equivalent to the *Crucifige*.[17] He had felt on him the brutal grip of the horse traders of the populace, and he believed he would faint for disgust and horror. "The spindly little Caesar, nervous, impressionable," said Vandal, "who always held in horror any physical contact with the masses, experiences a physical failure. His chest oppresses him, his sight is troubled, he has only a confused and indistinct perception of things anymore." He often spoke of his contempt for deliberating assemblies; he was not in the habit, and he made them see it clearly on that occasion. When he was taken out from among that rabble and he saw his troops again, he recovered immediately, understood his true role and that was the lightning. But his anguish had been

---

[17]*Crucifige*: Latin for "Crucify [him]." A reference to the Passion of Christ.

plenary, and he must have remembered it unto his dying day.

One has often said that life is a dream, and one knows the quasi-supernatural power of impressions that the soul receives in dreams. What to think about the Napoleonic dream that lasts twenty years, from Vendémiare to Waterloo? The dream of such a man, its effects on such a soul, and the ever-renewed anguish of such a dream!

There are popular images that have Napoleon sleeping on the evening before Austerlitz, at a moment when he had driven his admirable army and young empire into a corner, and when the least mistake would have meant irremissible disaster, as two hundred thousand Prussians were preparing to fall on him; even in the case of a victory it would not have been a triumph.

Those poor images are strangely significative. He slept under his "star," that naïve great man, but who could say whether that sleep was the repose of his soul? He had already had so many tragic hours of uncertainty, at Boulogne, at Marengo, at Verona, at Rivoli, at the Pyramids, at Saint-Jean-d'Acre, and which ought to have stayed with him to the end. Everywhere alone, in other words not having one among his lieutenants who was his equal and always forced to turn fifty thousand combatants, by the addition of his person, into one hundred fifty thousand soldiers; what must have been his secret anguish with each of his glorious steps!

The images do not say that he was sleeping on

the evening before each of his battles, but popular legend gives it to suppose, and that legend was right, at least on a deeply allegorical level. Napoleon was a sublime sleeper, a somnambulant victor whom others' and his own suffering made cry out during his sleep, and which cries brought fright to the ends of the earth. One day, he woke up again, without his sword, at the very moment when he appeared before God...

What a gulf of meditations, if one comes to realize that that immense man of war could never obtain a definitive victory; that after Austerlitz he needed Jena, Eylau, Friedland; and that then Wagram was necessary to bring him to that frightening dilemma that destiny presented: either to be vanquished in vain at Moscow or to be crushed elsewhere by the coalition of all peoples! Whatever you do here or there, your ruin is inevitable and you can do nothing about it. You are in the shackles of sleep, in the torture or the voluptuousness of dreams. A superior and completely infallible Will has decided that you will be the agitated spectator of your own incomparable life...

The sublime Tauler once said that the sky is in the human soul and that God is pleased to sojourn there. "Evil men also carry god inside themselves, but they would not know how to enter. And that there is the greatest agony of the damned, – to know that they have within themselves heaven and God, without ever being able to enjoy one or the other." I do not at all believe that Napoleon had been an evil man, and I believe even less that he had been among the damned, as affirm, with a silly emphasis, the imbecilic devotees or prostitutes of the Restoration. Paradise without my Emperor, – I do not conceive it. It suffices to see in him a man excessively superior to others, but, all the same, subject like them to the law of exile. Nobody can reintegrate with Heaven or the terrestrial Paradise of his soul from which the original Disobedience expulsed him. One would need to be able to bind all one's senses and leave them at the door, infinitely rare miracle to effect, obtained only by the saints whom the Church puts on its altars. Nonetheless, something of that sort happens sometimes during sleep, and it is for this reason that the impressions of joy, sorrow, or dread possess an energy then that is impossible to find again or to understand, when awakening has stirred the dragon of the senses again.

If one says that all Napoleon's life was a dream, it's enough to make one sweat fear thinking about the supernatural agitation of that Titan's sleep. At that time, all his battles would have taken place *in his soul*, and he would have regarded, or heard, them with an immense anguish, like a prodigious poem that Someone greater than, and more fearsome than, himself would have imagined.

Imagine now that there was, among so many other dreams, the dream of his Coronation and his Consecration by the Vicar of Jesus Christ, when he had all Europe shuddering and convulsed under the boots of his infantrymen, under the hooves of his innumerable cavalry, that there was, after miraculous victories, the nightmare of infinite disasters and the unimaginable apocalypse of his Return and his demise.

And all that on the threshold of his soul! No one who has never begged can understand anything of Napoleon's story.

He was, on the threshold of his soul, the Beggar of the Infinite, the ever anxious Beggar of his own end, which he was ignorant of, which he could not understand; the extraordinary and colossal Beggar asking for the small *sou* of world empire from whomever passed by, the distinguished favor to contemplate in himself the terrestrial Paradise of his own glory and who died, at the extremity of the earth, his hands empty and his heart broken, with the weight of many millions of agonies!

Chapter 4: The Battle

One pale day dawns over the sad plains of Poland. To the sound of the clarions responded the neighing of forty thousand horses. The cold and black night had weighed heavily on the army whose sleep had to be interrupted, how many times! by the near or distant moanings and groanings of the wounded from the day before and the day before that. Those lamentable cries traversed the memories or dreams of each and sundry, for each of his warriors possesses a soul that will leave his body in several hours probably. It is an immense drove of souls, it is Eternity's livestock.

Many, a great number doubtless, dreamt of their families, their fields, their villages, in Burgundy, Périgord, Normandy, Brittany; others, of those in Holland, Germany, Italy, and even Spain, for the Emperor's armies were recruited from everywhere, except Russia and England.

They fight for ten years, they will fight for another ten years surely, and nobody could say when nor how that will finish, Napoleon even less. The most intrepid leaders were already muttering. What they feel very acutely is this, that they have all Europe against them, simply because it is France, which is the living soul of all peoples, and that it is a law for the human brute to wage war against his soul.

For the humble soldiers that soul is visible in Napoleon, so visible that if he came to die, that would be the end of France and the end of the world. Is there anything more tragic, I ask you, than the tears of that

poor grenadier weeping at Berezina[18] for having seen
him marching among the specters of his Old Guard?
"In truth, I don't know if I'm awake or if I sleep. It
makes me weep to see our Emperor marching on foot,
a stick in hand, him who is so great, him who makes
us so proud!"

But that moment had not come. The peoples'
humiliation had not yet been sufficiently fecundated
and many more victories will be needed to give birth
to disasters.

While waiting, hear the preliminary din of the
artillery, the grandiose rumbling of the cannons. The
Grand Army stirs, stretching its powerful limbs,
yawning before death. To wake it up completely, the
freezing wind throws packets of snow into its face.
Now see it standing on its feet, shivering and shaking
in the valleys, on the hills, on the frozen lakes, in the
middle of the woods.

There are, here and there, on the chessboard of
the Infallible, the fearsome wildcats whom he dispos-
es: Davout, Augereau, Ney who knows neither fatigue nor fear; Murat the gutter of battalions, the
Achilles of all combats; the sublime Lannes, the
frightening cuirassier Hautpoul, the epic generals
Saint-Hilaire, Friant, Gudin, Morand, fifty others.
Fleet and precise like angels of war, they execute the
last orders of their master, and the carnage begins.

There must be, that evening, twenty thousand

---

[18]Berezina: a reference to the Battle of the Berezina, November
26-29, 1812, when Napoleon's army, in retreat from a disastrous
campaign in Russia, tried to cross, and eventually did cross, the
Berezina River, while the Russian army were at their heels.

dead and thirty thousand wounded at least, and there is no time to lose: for it is God who makes the Day of Man so that he might fill it with his good or bad works, and in February there are no more than eight hours in the day in that vicinity of the pole.

It is indispensable to have been a witness to one of those conflicts of multitudes to know how much life is like a dream. Imagine an entire division cut down by grapeshot. What does it matter and who would have time to weep? Thirty squadrons pushed by the Furies trample it underfoot in order to saber, a little farther, on the cannoneers and infantrymen, before succumbing themselves in the luminous night of the dead. Then the battle has its incessant ebbs and flows. A position taken with great effort is lost and reconquered, how many times! A heroic charge thought to be decisive has been stopped by a cyclone of gunfire; the half-destroyed cavalry are pulled back to the infantry who will protect them as best they can, having sometimes a furious need to be protected themselves. But the litter of the dead begins to weigh down on them, and the souls exited from the tomb of their bodies, the poor, previously tenebrous souls, knowing finally for what and for whom they have so savagely fought, have been transported away, invisibly, onto the imperial knoll, around the visible Master who divides them with his hand like importunate thoughts...

For victory is not his yet and he needs victory. Victory is his *Requiem*, the repose of his soul, is very own soul, in that obscure world. It is his bread and wine, it is his resting place, and it is his lamp. Was he created for anything else than victory then? When one of his corps falls back, it is as if he is physically beaten by the horse crops, by a multitudinal thrust. But his face is as impassive as bronze which lets nothing of his torment show through. Perhaps even he does not suffer, so strong is his heart, so great the impavidity of his genius! He will suffer later, without a doubt. But at this moment he appears happy, he feels his power. He knows that he is the guardian of Fortune's abortions, he has the *arcs-de-triomphe* for Incertitude and even for eventual disasters, perfectly certain of always finding from deep inside himself some unforeseen and stunning resource that will make him even more powerful.

At that moment he looks, once again, at his battlefield and, tranquilly, "he takes three steps forward, like the Gods." Among all his profound schemes, ineffectual until now, suddenly appears a Maneuver that reminds one of the infant Hercules spattering all the heavens with the milk of Jupiter's wife. Murat has just passed like a torrent, crushing all Europe before him, in half an hour, in four square kilometers, and Napoleon needs his soldiers to take only several more several steps for him to become Emperor of the Occident.

"The fate of a battle," he said on Saint Helena, "is the outcome of an instant, of a thought. One approaches it with divers schemes, one gets mixed up in it, one fights for a certain amount of time; the decisive moment presents itself, a *mental flash* speaks, and the smallest reserve accomplishes it."

He acknowledges that he was very profoundly

moved by the scene on the fields of Eylau, so red with blood that the snow must have been tainted with it until the end of winter. Impossible to doubt that emotion when one has studied Napoleon. He is more man than other men by virtue of his infinite superiority. But that very superiority "tethers him to the shore" of an impassivity necessary to his prestige. "One particularity," says Thiers, "stands out in everyone's eyes. Either tending to return to the things of the past, or also for economical reasons, an idea was floated to make the troops' uniforms white. One had tried it out with several regiments, but the sight of the blood on their clothing decided the matter. Napoleon was filled with disgust and horror and declared that he wanted nothing other than blue uniforms, even if it cost more." He could not help, in spite of everything, from betraying, on that occasion, the distress within his heart, in one of those terse and fateful Bulletins that he used to rattle the world with.

For those who see into the Absolute, war makes no sense unless it is *exterminating*, and the very near future will demonstrate that to us. It is a stupidity or a hypocrisy to take prisoners. Assuredly, Napoleon was neither a sot nor a hypocrite, but that supposed executioner was a *sentimental* person, always ready to forgive, a magnanimous person, believing all the same in the magnanimity of others, and one knows what that cost him, that incomprehensible illusion.

At Austerlitz, he let Alexander go free, whom he could have taken prisoner; after Jena, he left the throne to the battered house of Prussia; after Wagram he neglected to break apart the Austrian monarchy, etc. Finally, at Rochefort, he trusted in the generosity of England! He knew the horror of the *bagnes*[19]; it would have been easy for him to make reprisals while sending to hard labor, not the poor sailors or soldiers, but an entire elite of English society in France after the rupture of the peace of Amiens, terrible expedient and which would have probably been more efficacious than the continental Blockade. He reproached himself later for not having done it and for having thus lacked character...

He was not therefore the monster that would have been required for total, apocalyptic war, with all its consequences, the abyss of war invoked by the abyss of turpitude, and he is evidently not that demon whom he will have been the precursor of.

---

[19]*bagnes*: galleys, or dry-dock prison ships.

Chapter 5: The Globe

In his excellent book, *Napoleon and Alexander I*, Vandal, recounting the Napoleonic ceremonial at Dresden, in 1812, wrote this:

> *In the evening, the sovereigns got together for dinner, which took place at the Emperor of the French's place. They got together beforehand in his apartments. There, if one must believe the story, in his manner of operating and having himself announced, Napoleon affected a grandiose simplicity that set him apart from all the powers gathered to hear him and which elevated him above them. His guests were announced by their titles and positions. There were, to begin with, the innumerable Excellencies and Highnesses, Highnesses of every region and every provenance, ancient or recent, Royals or Most Serenes, – then the Majesties: Their Majesties the King and Queen of Saxony, Their Imperial Majesties and Royal Apostolics, Her Imperial Majesty of the French, Queen of Italy. When all those sonorous appellations had echoed through the rooms, the august assembly found that it was complete and the Master could arrive. Then, after a small interval of time, both sides of the*

> *door opened again and the usher an-*
> *nounced simply: 'The EMPEROR!'*

Twenty-eight years later, at the Hotel des Invalides,
one waited for the remains brought back from Saint
Helena. No majesties or highnesses had been announced. Why would they have come, having nothing
more to beg or hope for from that Dead Man, the victim formerly of their cowardice or their perfidy?
Since May 5, 1821, his Europe, moreover, was no
longer recognizable. The infamous and ridiculous
Bourbons of the so-called eldest branch, supplanters
of his glory, were abhorred. The crown-bearers who
were his contemporaries, or his domestics, squatted
on earth or under it, and nothing since had been accomplished in the world that merited any attention.
The rare and rarer debris of his Grand Army were ill-
persuaded of his demise and latched on to the hope of
another kind of return. One cannot do entirely without
Beauty, and it was really too ignoble to subsist in the
legitimate or illegitimate filth swept from all over Europe onto poor France since 1815. From the entire
ceremony of 1812, with the transitory elements having been trimmed away, there remained merely this:

The door opened wide finally, having remained closed for a long period of time, before a
panting and silent multitude, and the grave voice of a
veteran of Wagram or Moscow announced these simple words: "The EMPEROR!" It is said that several people fainted for enthusiasm on seeing the coffin enter.

It seems to me that that is greater than Dresden, and that that last triumph is incomparable. What
was returning then was not only the infinitely pre-

cious relics of a man whose greatness had appeared to equal that of a saint, it was the imperial Globe in the hand of its Master, who had been the soul of France more than any other hero or prince of any other period in its history whatsoever.

I said earlier that such was the deep feeling of his soldiers. When those poor people died while crying: "*Vive l'Empereur!*" they truly believed in dying for France and they were not wrong. They absolutely died for France; they gave their life just like that, as had never been done before, not for geographical territory, but for an adored Leader who was in their eyes the Country itself, the undelimited fatherland, as unlimited, resplendent, and sublime as the great valley of Heaven, which no scientist could tell the frontiers of. There was India, and there was immense Asia, the Orient after the Occident, the Globe really, of the universal Empire, in the terrible claws of the Roman Bird domesticated by their Emperor, and their Emperor, he was France, – equivocal, enigmatic, indiscernible before its apparition – from then on precise in its majesty, irradiant and clear as day, God's young France, the France of good bread and good wine, the France of glory, immolation, heroic generosity, immeasurable grandeur, all the litanies of heart and thought!

Stat Crux dum volvitur orbis.[20] That was it really: Napoleon having replanted, on this old globe, which had become his, the demoralized Cross. *Volvitur*. Where had it not turned since Charlemagne? How

---

[20]*Stat... orbis*: Latin for "The Cross stands [still] while the world turns [around it]." It is the motto of the Carthusian Order.

many had there been who believed they held it in
their hands full of dust?

After Louis IV, the so-called "*Child*," who
was, in Germany, the last Carolingian, there had been
the very illustrious House of Saxony, the three magnanimous Ottos, and Henry the Saint, the flower of
the Middle Ages in its spring, and then the cohorts of
the Houses of Franconia and Swabia. There were the
chiefs of what was called the Holy Roman Empire
come from Holland, Cornwall, Castille even, Nassau,
Austria, Moravia, Luxembourg and Bavaria. There
were finally the Hapsburgs from whom Napoleon was
to seize that magnificent symbol of domination, become by them an emblem of sterility and turpitude.

What could it mean, moreover, in the Ghi-
belline hands of those Germans, the venerated simu-
lacrum of the Christian omnipotence of Constantines
and Theodosiuses? A Napoleon was required to restore it, in his person, to the Latin World, so long fall-
en into the mud. In his person and forever. The imperial Globe is forever now in the great Tomb of the Invalides where there is room for only one deceased.
Nobody will come to take it up again, even if that
man had ten million men behind him.

"The desert," said Las Cases, "had always
held particular attraction for the Emperor... He took
pleasure in observing that the name Napoleon means
Lion of the desert." In what language? I don't know.
But it is quite certain that that mirage of his imagination is a profound reality. He himself was the desert,
creating around him, living or dead, so vast a desert
that all men on earth could not fill it, and their multi-

tudes would seem nothing compared to it, in the eyes
of God, in the silence of space.

Chapter 6: The Bees

*May 27, 1653, near Tournai, in that
part of the Low Countries that France,
for so long, envied of Spain, the au-
thentic tomb of Childeric I was found.
The magistrates took great pains to
take possession of the objects that
their assistants had rapidly lifted a
portion of already. Of the two hundred
rare jewels that had been discovered
at the time of the excavations, only
about thirty remained. They were
golden bees, their wings garnished
with red glass, mounted in compart-
ments. The small metal ring that some
of them preserved must have been at-
tached to a piece of fabric. A scholar
declared that they had ornamented the
king's mantle, maintaining that the*
fleurs-de-lys *of the French blazon
would have been a distortion of those
bees. Now, Napoleon I, who loved to
speak of his most distant predecessors
and who wanted, the day his imperial
eagles were distributed in Boulogne,
to sit on Dagobert's throne, became
interested in Childeric's relics. By his
orders, the bees from the tomb in
Tournai were imitated in order to replace the* fleurs-de-lys *on the Sacred
Imperial Mantle, which had decorated*

the Capetian kings' mantles. Remark-
able fortune of that Merovingian vest-
ment.[21]

After fourteen centuries, there is not a lot that
can be said about that father of Clovis, who was Childeric I. All that we know about him is that he scandal-
ized the Franks "by his lust," which could not have
been easy, and that those chaste barbarians, after having expelled him for a period of time, replaced him
with the Roman general Ægidius. One also knows,
according to the good Saint Gregory of Tours, that the
Queen Basine married him "for his merit and great
courage."

Dagobert is without a doubt more interesting,
and one can understand why Napoleon would have
the desire to sit on the great Merovingian's uncom-
fortable and millenarian throne. But Childeric was
very dear, in his eyes, having been, almost, the most
ancient king of France, and also for having been
found in his tomb with the golden bees which were
mixed together with his very ancient dust. There was
also this, most certainly, that the bees must have suited his Latin soul, much more Virgilian fundamentally
than Corneillian, in spite of his decided taste for tragic drapery.

Saint Bernard compared, I believe, with more
charm than profundity, Jesus Christ as king to a bee
"having the honey of mercy and the stinger of justice." But Saint Bernard wasn't anticipating Napoleon, and Napoleon, assuredly, never read Bernard.

[21]*Revue Napoléonienne*, January-February 1912. The Boulogne
mentioned may refer to Boulogne-Sur-Mer.

The famous parable of Samson and the Lion, feebly echoed in the fable of the bulls of Aristaeus,[22] suited him better and was, I believe, more likely to be known by him.

Whatever the case, Merovech's[23] bees pleased him and he wore them on his shoulders, while he set the world on fire, until the day when those flies, irritated finally with their master and as traitorous as men, transpierced him. They died, it is true, at the same time he did, and the same experience attempted by his nephew, six lusters later, did not seem any less lethal.

For it is a terrible danger to touch symbols. "Guess or I will devour you," they seemed to say, like the Sphinx, to the travelers bold enough to venture along the route to Thebes, enigmatic capital of Boeotia. It is a road to be avoided when one is not, unlike the first Napoleon, invincibly driven.

God forfend that I should attempt an explanation. The imperial mantle's bees are as mysterious to me as they must have been to the dusty Childeric I and to Napoleon himself, as perfectly undivinable as Solomon's enigmas or the Gospel's parables. It suffices to hope with certainty that we will know someday what they meant in the destiny of a great Emperor and in that of our old world, which does not stop descending into darkness until finally it disappears.

---

[22]Aristaeus: a Greek demi-god credited with having discovered apiculture, among other useful arts for man.

[23]Merovech: the Salian-Frank king (reign AD 411-458), father of Childeric I, and founder of the Merovingian dynasty of France.

Chapter 7: The Footstool

"The earth is a man," said I don't know which mystical philosopher. That strange phrase comes back to me suddenly as I think, once again, on the imperial Globe that I see ever rushing from time immemorial to consign itself into Napoleon's hands finally. That globe naturally expresses the terrestrial sphere, reverse image of the celestial sphere where it appears to be nothing more than an imperceptible dot. But Space as well as Quantity is merely an illusion in our minds. A Number is merely an indefinite multiplication of the primordial Oneness and nothing more. It is therefore probable and even certain that the minuscule earth, so vast for the poor men forced to circumnavigate it, is, in reality, greater than everything, given that God incarnated himself to save even astronomers.

That Incarnation is not just a Mystery, as we are taught, it is the center of all mysteries. *Omnia in* IPSA *constant*.[24] When one reads that the Son of God, his Verb, "was made flesh," it is exactly as if one were reading that he had been made *earth*, given that the earth is the substance of the flesh of man. But God, assuming human nature, acted necessarily according to his divine nature, in other words, in an *absolute* manner, becoming in that way more man than all the men formed on earth, becoming the Earth himself in the most mysterious, the most profound sense.

---

[24]*Omnia... constant*: Latin for "Everything consists in it." Possibly a reference to Colossians, 1:17: "*et ipse est ante omnes et omnia in ipso constant*."

When one mentions the earth, it is therefore the Son of God, the Christ Jesus himself that one refers to, and it is enough to discourage every exegetic constancy to discover that the word *terra* is written much more than two thousand times in the Vulgate, to say nothing of the word *humus*, invocator and synonym of *homo*, which one can read exactly forty-five times there.

Filled with these thoughts, open the Holy Book and you have something like the tearing of the veil over the Abyss. You will be immediately the bowled-over witness to the nuptials of Rapture and Terror. You will no longer know anything, you will no longer dare to speak. You will no longer dare to spit on the earth which is the face of Jesus Christ because you will feel that that is really how it is. When you read, for example, in Saint John that Jesus "wrote in the earth with his finger," in the presence of the Scribes and the Pharisees accusing his Bride, the Church, for whom he had to die, to have been "surprised in the act of adultery," you may feel, with an unknown emotion, that that Redeemer wrote *on his own Face*, with the same finger that he had used to heal the blind and the deaf, the silent condemnation of implacables and imbeciles. "He who is issued from the earth is of the earth and speaks of the earth," had said his Precursor, and it is for that reason that the Master always expressed himself in parables and similes. There would be no end to it if one had to unwind, with a trembling hand and with a heart beating like the bells of the Epiphany, all the concordances in the Holy Text.

So a boundless respect would be needed for this miraculous earth, inexpressibly sullied by all peoples for so many centuries and so cruelly dishonored today by greedy industries that despoil it of all its décor, after having violated it to its entrails. But all the malice of demons will not insult it any more than that Face of the Redeemer was insulted. Try as one might to sell it or exchange it unjustly and by the most ignoble roundabout means of cupidity, it will never add up to an equivalent quality of outrages. However devastated the visible face of our globe might be, one will not despoil the hidden treasures of anger of Him in whose image it is, and one will never extinguish the immense furnace of his heart either.

"When I am raised from the earth," said the Master, "I will attract everything to me." One has wished that that spiritual prediction was fulfilled in the visible world; and that those who were, for a time, the sovereigns of this world, without realizing what they held in their dirty hands, had planted the Cross on *their* globe in order to attract everything to them. That was the secular deception until the arrival of Napoleon, who had to be, by divine decree, the last and the loftiest victim.

Nothing like him exists nor can exist anymore, the only being in whom all things appeared, for one moment, to have their consistency, Napoleon the Great having been swallowed up in his turn. Never has a king or emperor fixed so penetrating, so attentive a gaze on the earth. Imagining perhaps that it resembled him, with its volcanoes and its oceans, he considered its distress, the horror of its wounds, its bruises, its scars, its mortal lividity, observed even the beginning of its agony. Physician more than temerarious man, he undertook to heal it, to renew that moribund face by infusing it with a new life. He succeeded merely to cover it in blood, and that was doubtless the only thing he could have done, given it appeared to have profited by those terrible treatments. Even after one hundred years it has not finished dying. Its obsequies have been canceled, but the new Cross that Napoleon had given it crumbles to dust, and the very idea of a Globe falls apart, the sphericity of the earth being contested by scientists who attribute to it I don't know what other geometric shape.

When will he come, Him who must come and who was, under Napoleon, merely foreboded by the universal trembling of peoples. He will come, clearly, in France, as is appropriate, Our Lady of Compassion having wept at La Salette while speaking about Him... He will come either for God or against God, one has no idea. But he will certainly be the Man expected by both good and evil people. Supernatural missionary of joy and despair whom so many prophets have announced, whom the cries of fearful or fierce beasts have prefigured, as well as the birds' limpid or melancholic song, the clamor of gulfs or fearsome exhalation of charnel houses, – since the time of the Disobedience of the Patriarch of Humanity.

On that day then one will know finally the true *shape* of the earth and why it is called the footstool for the Lord's Feet.[25]

---

[25]footstool for the Lord's Feet: a reference to Matthew 3:35.

Chapter 8: The Tiara

"We do expect to go as far as the gates of hell, but we intend to stop there." It was in those terms that the dolorous Pius VII spoke of the Concordat of 1801, terrible stipulation when necessity constrained him not to let the poor flame of that last light of the world, which was France, go out completely.

It was necessary even, the repugnance was so great, that Napoleon should take a supernatural ascendancy over that gentle and timid old pontiff, who appeared to see in him a bit more than a man, and whose worst treatments were not enough to discourage him in his affection. For the bewitching power of that conqueror is the occasion of an astonishment as never to be seen again. It was quite simply that he was adored by his soldiers whose courage he centupled and whom he associated daily with the most inundating glory. It was the most natural thing in the world that the ministers of his power, the innumerable functionaries of his empire, were amazed by the number of prodigies they saw him effect. The sovereigns themselves, his adversaries or his rivals, so often vanquished and humiliated, could not help admiring him and trembling. A million hands went out from him, in order to take souls.

But the Vicar of Jesus Christ, was that possible? Supreme Doctor and Pontiff, infinitely elevated above all men, not by nature or culture, but by magistery and ordonnance of God; Primacy of honor and jurisdiction in the universal Church, Cornerstone and

Bearer of the Keys, without superior or equal on earth; infallible and sublime Judge unable to be judged himself or deposed by anyone; is it credible that Pius VII, nowise unworthy successor of so many saintly Popes, could not evade that prestige? And nevertheless that is certain. Pius VII loved Napoleon with a love of predilection that placed him in his heart above all other princes, to the point of incurring the reproach of *partiality*, practicing thus a sort of nepotism in favor of the conqueror of the universe, as if he had been his closest son. Even when he had to suffer from that hand and suffer to the point of agony, his tenderness for the prodigy appeared to augment. However, the Emperor could not make him prevaricate, not even in form, in 1813, at the time of that false and surreptitious concordat signed by the distraught, almost moribund, septuagenarian, who came back to him, immediately after; concordat of null worth and subsisting in the papers of history merely as a proof of the moral violence exercised by Napoleon against his captive.

"We have done everything," the Pope had said, in 1807, before the rupture, "to ensure good communication and concord; We have arranged to continue to do so for the future, provided that one maintain integrity of principles with respect to which We are *unwilling to budge*. It is a matter of Our conscience and on that no one will obtain anything from Us, even if We should be flayed, *ancor chè ci scorticassero*."[26] That so simple firmness exasperated the emperor who became, for a moment, a prophet

---

[26] *ancor chè ci scorticassero*: Italian for "even if he should flay us."

against himself. There was a threat of excommunication. "Excommunicate me?" Napoleon wrote, July 22, to the viceroy of Italy, "does Pius VII think the hands of my soldiers will fall off?" Exactly five years three months were needed before October 1812 came.

The great soldier wanted world empire so badly that his intelligence had been clouded to the point that he no longer understood that an order may no more be transgressed by a pope than by a grenadier, and that there are things that cannot be demanded. "The Pope reigns over the spirit, and I reign over matter," he cried in his despair. "*The priests keep the soul and throw me the cadaver.*" What flashes in the night for that great man, and how many in vain! He persisted in misjudging the point at which the need for force must stop. Could he ignore however that; in the natural order of things, power acting to excess itself creates and finds in the end a *resistance* that it can no longer vanquish? Absurdly frightened by imagined attacks by the Holy See, which was merely defending itself, Napoleon made the deplorable decision to abduct. The Pope, although profoundly miserable to have to punish, responded by the excommunication which he retracted a little later, when divine protection appeared to have been withdrawn from his enemy whom that formidable sentence prevented, it is said, from sleeping.

There had been other pontificates as troubled as Pius VII's, but none could bring its titular to so plenary a feeling of bitterness. The cross inflicted by Napoleon was incomparably harder and heavier than all others. It was the cross of genius, the cross of heroism, the cross of military glory that had never seen an equal, the cross of unbounded human grandeur, the cross of all terrestrial prefiguration, the *cross of honor*! The unfortunate Pontiff, crushed beforehand by the weight of the Keys, had to carry that burden as well. He had to carry it for fifteen years, and it is a miracle that he didn't succumb to it.

His immediate predecessor, Pius VI, the Pope of the Revolution, had had a very hard life, and he was forced to die in exile, *not far from La Salette*, having seen the entire ancient world crumble around him. Long before the revolution broke out, it was already a torment to govern the Christian universe. "Alas!" said Pius VII, Pope during the Consulate and the Empire, "We have true peace and true repose only in the government of Catholics subject to infidelities and heresies. The Catholics of Russia, England, Prussia, or the Levant cause Us no trouble. They ask for bulls, guidance for which they have need, and they march, after that, in the most tranquil way following the laws of the Church. You are familiar with all that our predecessor had to suffer from the changes effected by our emperors Joseph and Léopold. You are witness to the assaults against Us everyday made by the courts of Spain and Naples. There is nothing so miserable today as being the Sovereign Pontiff. He is guardian of the laws of Religion, he is its Supreme Leader; Religion is an edifice that one wants to attack from all sides saying that one respects it. One believes one has need of Us to operate continual subversions, and one does not consider that it is Our conscience and Our honor that refuses all those changes. One rejects our objections with fits of temper, with anger;

requests come to Us almost daily accompanied by threats." And the French ambassador, the spiritual Cacault, reporting those grievances in a dispatch to the First Consul, added boldly: "No fetish has been so beaten and so maltreated by its negro as the Holy See, the Pope, and the Sacred College of Cardinals have been, for ten years now, by Catholic believers."

But what were all the prior harassments and squabbles, going back to François I or earlier, compared to the zeal of that "devout son" Napoleon when writing to the Pope, on February 1806, the unprecedented letter in which he declares himself to be Emperor of Rome and which could be summarized as follows: "I concern myself more about religion than you yourself do; you leave it hanging, watch what I do; I will be wiser, more capable, more pious even than you, who let *souls perish*." (!!!)

The consuming activity of that soldier who knew nothing about the government of the Church, could not admit nor conceive of the slowness of Roman decisions, and a furious impatience agitated him as often on the throne as in the field. Pius VII tried in vain to explain to him that velocity in ecclesiastical affairs is a symptom of corrupt practices. Soon they no longer had means to understand each other, no lasting accord proving possible between those two men, the one holding an immense Sword, but for one day only, the other presenting divine Law, changeless and without end.

It was in very good faith that in the beginning of his grandeur, Napoleon wanted to heal the wounds of the Church, of the whole Earth, as I have already

said, and that in 1806 and later, he still meant to. But
the Absolute is incompatible and the absolute that
was in the Emperor's will would never be able to
make the Keys of the Holy Ark turn there where, under the Pope's eye, the divine Will resided.

The first serious difference of opinion is the
refusal to abrogate Prince Jerome's Protestant marriage. On that occasion Pius VII took pains, highly in
vain with respect to his effort, to write a long letter of
angelic serenity, worthy on all points of the most holy
Doctors. A little later, there's the occupation of An-
cona, in contempt of pontifical neutrality, first symptom of the rage of dispossession. The Pope complains
about that injustice with an apostolic and paternal
gentleness which had no other effect than to harden
that Pharaoh's heart. At that point, there was nothing
left to do. The Church, deprived of its leader, is
forced to wait, while suffering and moaning, for the
great victor to succumb.

The prodigious man of Jena and Lobau, who
had need of his continental Blockade in order to prefigure the Devil or the Holy Spirit, went, the bottom
of his heart not being there for nothing perhaps, to
that extremity of oppression where it becomes inevitable that the floodgates of heaven should break.
"*It is prohibited* that the Pope Pius VII communicate
with any church in the Empire, *under pain of disobedience.*" That political counter-excommunication, so
similar to an injunction of the police, was delivered to
his Captive on January 14, 1811.

The following March 19, immensely remark-
able date, the King of Rome was born. The Patriarch

of Obedience, whose feast it was, and whom another Pope had proclaimed as the Patron of the universal Church, received then in his arms that poor child from the greatest of men, and, as he was also the Patron of good death, he returned him as soon as he could to his true father, the Emperor of worlds.

In 1809, a few days after his abduction, Pius VII, being dragged from city to city, passed through Grenoble. There, the two remaining insurmountable resistances that Napoleon found on the continent, the Holy See and Spain, came together. The prisoners of Saragossa were in Grenoble. On the arrival of the Head of the Church, everyone rushed forward, kneeled at his feet, and the entire city imitated them. Napoleon, on the Danube at that time, sensed perhaps a cloud passing over his head. His "star" went pale. One had, for some time, stopped noticing it at Baylen and at Sintra; it had almost gone out at Essling, that strange star which would have led him to Bethlehem maybe, if only he knew how to kneel one single time like those he had vanquished, and which led him instead to Saint Helena, – the mother of Constantine having prepared for him there a solitary tomb where the cross of hope,[27] granted to the least of castaways from the Ocean, was not at all permitted him.

All that seems, today, incredibly distant. The judgments of men have replaced their angers, but one still does not see, among historians, a superior discernment into the magnificent events of the first Empire. Nobody was aware of it until something tran-

---

[27]mother of Constantine: Saint Helena, the mother of Constantine I, whose feast day is August 18 in the Roman Catholic Church.

spired between the two greatest powers, the *only* in reality, God and Caesar, – something ineffable, and comparable only to the parables or prophetic prefigurations of the Old Testament repercussed mysteriously on all the pages of the New.

Now the heart and voice falter. No one knows anymore what to say or what not to say. Here is, for example, Moses, the huge Leader of God's People, to whom the Lord "spoke face to face, like a man accustomed to speaking with his friend." In punishment for his complaints, God's People are cruelly afflicted. Moses prays and the Lord commands him to erect a bronze serpent, the mere sight of which will heal all those who look at it. That serpent would signify then both the ancient Enemy of men and their Savior; it is the symbol of the Tempter on the Cross of Redemption, and he who institutes that frightening and salutary Sign is the obedient Vicar of God in the desert, the incontestable predecessor of the Vicar of Jesus Christ, in those distant times. Would that not be then, – I can barely write it, – at that distance of forty centuries, a marvelous analogic symbol of the CORONATION of Napoleon by Pius VII, coronation of an *usurper* so often compared to the Anti-Christ, so that there might be presented to the expiring world a sign just like the hope of miraculous healing? With a little boldness, one could go so far as to say that that coronation for which the very gentle Pontiff was blamed so much was perhaps, in the thought of that confidant of divine Charity, like Extreme Unction administered to a very ill Europe, condemned by the most knowledgeable of physicians.

Finally there are those two Souls: the *central* and immoderate soul of that unique Napoleon, on the one side; on the other, the soul of the imperishable Papacy. Who would think then and who would dare sustain, after one hundred years, that there was really any antagonism? God had wanted Napoleon, as he had wanted all the popes, as he had wanted his Church. They needed to get along together and in a certain accord, at whatever price it might be; the one to dig a deep chasm between the old world and the new, the other to say to all peoples:

"Behold the *Delimitator*! His hand is hard and his foot heavy; but He whom I represent wanted him to be like that and not otherwise. If I suffer because of him that will be in the infinite and perdurable certainty of having done what needed to be done, at such and such a moment, for God and men. If that predestined man breaks me, it will not be without his having first unrooted himself. But the Tiara that I have the honor of wearing after so many others, will not be broken. Recognize therefore in him and in me the Will of the heavenly Father being accomplished at the same time here on earth as it is in heaven."

Chapter 9: The Canker

Napoleon, on Saint Helena, had condemned his enter-
prise in Spain. "That unfortunate war ruined me, it di-
vided my forces, attacked my moral standing in Europe. I began the affair quite badly, I have to admit;
the immorality must have appeared too patent, the injustice too cynical, and everything remained extremely nasty, given I succumbed. For the attack shows itself in its hideous nakedness for what it is, deprived
of all the grandiosity and numerous benefits that filled
my intention... Bayonne was not an ambush, but an
immense coup d'état... I dared to strike too high. I
wanted to act like Providence."

Like Providence! That's pure Napoleon. Feeling himself confusedly called on to prefigure Him
who must renew the face of the earth, he thought himself designated to operate that renewal himself and
many believed it with him. That is how he could be,
for ten years, the arbiter and kneader of Europe.
Without that prejudice, his marvelous battles would
not have sufficed. But there was Spain, which didn't
want to let itself be kneaded, and the Cromwell of European monarchies met his grain of sand in the ureter
of the old world.

That Spain of granite and guitars was a
strange place that had much to expiate. Unfaithful to
its mission of Christianizing America, it had fero-
ciously destroyed entire peoples. The iniquitous gold
of its galleons of torture and despair had, for a long
time, rotted its heart and liquified its brain. Its

Catholic kings, the richest on earth, it is said, were there, like the ridiculous sun of the Bourbons, ruling over several millions of superb beggars gnawed at by vermin. Religion, decanted from the sublime hearts of Saint Theresa and Saint Jean of the Cross into voluptuous or savage souls aching by the fetishism of the most material devotion, had become hideous.

No contact with any other people except, by force, with detested Portugal which barred it from the Atlantic, preventing it from perceiving, on the other side of the ocean, the Continent of gold. Deprived forever, since Utrecht, of its ancient possessions in Italy and the Low Lands; the recluse behind the Pyrenees whom Louis XIV thought he had destroyed; harshly spurred in Gibraltar by heretical England; that dominatrix of half the globe, two centuries earlier, subsisted from then on like a wild and poor woman, unapproachable on the checkerboard of its mountains where new ideas did not penetrate. In the cities, there were still, here and there, some men capable of seeing that their monarchy was garbage and feeling that something new was beginning. They paid, however, a very high price for that clairvoyance, having had their throats cut quite inhumanely by their own citizens, from day one. But the people of the countrysides saw nothing and felt nothing, except that they were going to be treated maybe like their ancestors had treated the aboriginals in the New World so vainly confided to the charity of Catholic Spain by the very gentle Messenger of the Redeemer, Christopher Columbus. Then it was a war of demons.

There was however a very marked difference,

and I ask all Spaniards their permission to express it. French soldiers, in the beginning and when the welcome by stabbings of the knife had not yet enraged them, were really the naïve disillusioned ones of '89, thinking they were bringing deliverance everywhere and fraternizing with all peoples, as stupid an illusion as one might wish for, but certainly generous, which it is equitable to oppose to the prickly individualism of Spain, as closed as China to every foreign interference and profoundly indifferent to the adversity as well as to the prosperity of other inhabitants of the globe.

From 1808 to 1814, they were massacred, they were infernally tortured, and that war could not end except at the end of the great empire. Three hundred thousand French troops, thrown by Napoleon on that miserable kingdom awarded by him to an imbecilic brother, worked their way across it in every direction, destroying men and things, burning, pillaging, slaying, raping and profaning, in reprisals of most dreadful cruelty. More than two hundred thousand Spanish combatants remained there, and how many of the emperor's soldiers came home again? The known numbers are enough to make a man tremble. At Saragossa alone, a report by Marshal Lannes records, with horror, more than sixty thousand dead enemies!...

One has often asked why the great conqueror, available after Wagram, did not return to Spain to finish the war. It is quite certain that Wellington could not hold the line before him and that then he would not have had need to cross over the Nieman or go to

Moscow. But that, that's the mystery, at each instant met in the life of Napoleon. Obeying his implacable destiny as prototype or paragon, it was necessary for the monster of activity to become inert at that moment in order to fulfill the punishment of both one and the other. It was necessary for him to consummate his appalling Austrian marriage and in that way to make sure of a rupture with the Barbarians of the north.

The ignominious capitulation of Baylen had taken place in the vicinity of Las Navas de Tolosa, glorious battlefield for the Spanish for five centuries nearly, and one knows how much that unhoped-for triumph exalted them. That was the first blow. Europe understood that the colossus, no longer appearing invincible, was shaken, and he himself felt that the earth was tired of carrying him. His all-powerfulness, although granted from on high, was so human, so fragile! How could he not have seen it? Assuredly, he did not know that he was an instrument, nothing more than a magnificent instrument, for the monstrance of a divine parable. All the same, he must have had the intuition of an initial redoubtable warning and the discussions at Erfurt, immediately afterwards, that "flowerbed of kings," as it was called, must not have inspired him much.

His one appearance in Spain, if inopportunely abridged by the armaments of Austria, had come to nothing. Conquest of that wretched peninsula was trusted to incompetent or disloyal lieutenants who did not know how, or never wanted to learn how, to work together and who, moreover, had always been condemned to insuccess, in advance, by the surprising incompetence of a fictive king. It was the poor soldiers who were forced to pay – appallingly.

One has spoken a lot about the patriotism of the Spaniards, of the *awakening of a people*. What has not been spouted about that commonplace? It is as if one were speaking about the patriotism of the Vendeans who fought solely for their priests.[28] What connection could exist, in that essentially provincial and *parochial* nation, between the savage peasants of the Manche or the toreros of Andalusia and the *montagnards* of Asturias, for example, or the unsociable herdsmen of Aragon! Nothing else, doubtless, than strict and frenzied religion, but identical everywhere, – that they liked their Capuchins and their curates. It was enough to eternalize a diabolical war. If Napoleon understood nothing of that deep character of Spain, what could those unfortunate soldiers, raised in ignorance or contempt of any religious practice have understood?

The conqueror of kings, habituated up until then to receiving the keys of empires or capitals, after decisive victories, was surprised by a people incapable of capitulating, forever elusive and wanting nothing to do with the war except perpetual ambush and the continual exchange of atrocities. That evidence disgusted him, and he let things go as they could, hoping maybe for lassitude, sacrificing thus half his fine armies, trying to forget the horrible

---

[28]Vendeans who fought...: a reference to the counter-revolutionary movement in the Vendée, similar to the Chouannerie, during the French Revolution.

wound on his feet in order merely to dream of the crown of all Caesars that he thought to strengthen around his head on fire. There is not in all history a more sorrowful page. The inexpressible calamities that came afterwards did not have that same aspect of tragic evil, that abominable aspect of sanguinary disloyalty and fratricidal fury...

"That canker of Spain from which there was no turning back," said the dying and captive Emperor, "... that fatal war with Russia, that horrifying rigor of the elements... and then the entire universe against me!... O *destiny* of men!"

Chapter 10: The Vile Island

"England traffics in everything," said Napoleon, with a bitter good-naturedness, the august prisoner of Lord Bathurst and Hudson Lowe; "why doesn't it start selling freedom?" One has to believe that he was feeling the absence of that merchandise and that he would be feeling its absence forever.

Why did one speak of English freedom? Another classic commonplace if ever there was one. And what nation is more a slave to its religious or political prejudices, its institutions, its diabolical pharisaism, its insurmountable and pitiless pride? Let's speak rather of the freedom of Carthage where one crucified lions, that is to say, its citizens who disdained commerce, or the freedom of Rome where insoluble debtors became, by virtue of the laws, slaves of their creditors. The Roman hypocrisy, which could only be surpassed by Britannic hypocrisy, had built a temple to Freedom on Mount Aventine. There they deposited the State archives. The Goddess was represented there as a woman dressed in white, symbol of innocence, having a cat at her feet, animal inimical to all constraint. England had replaced that perfidious feline with a leopard, – and that is, more or less, all the difference.

To the government of dynastic interests, dominant preoccupation of the kings of France and above all of Louis XIV, molecular predecessor of Napoleon,

is opposed, in that nation – as modern in the baseness
of its covetousness as it is ancient in its harshness
with respect to the weak – the exclusive government
of mercantile interests. For such is the shame and the
indelible tare of England. It is a Carthaginian usurer,
a trader of articles of public toilette, its insular isola-
tion allowing it, said Montesquieu, "to insult everywhere" and to rob with impunity. The famous tradi-
tional Rivalry is nothing else than the secular antagonism of a noble people and an ignoble people, the ha-
tred of a greedy nation for a generous nation.

"The idea of annihilating England," Sorel re-
marked, "was an idea that existed at the end of the
ancien regime in France, it was found to be simple
and natural, one discussed it seriously. The archives
are full of plans of raid." Napoleon thought and said
that nature had made Great Britain one of our isles.
At Boulogne, doubtless, he saw it cut up into forty
French departments, with an eventual autonomy for
Ireland and perhaps Scotland. His plan of invasion
was quite near to succeeding, and England, which
was dying with fear, become magically prodigious,
and rushed to throw itself behind the Austrian and
Russian armies.

For that old beggar woman, *Old England*, for
want of a young Empire that it could force to get
down on its knees, was reduced to offering itself, easily, as a consoler or middle-aged procurer on the
brink of ruin. Nothing but money was talked about
anymore; Europe became a market for human blood
where the Buyer was often deceived in the quality of
the globules or the quantity of its effusion. The decep-

tive peace of Amiens had been merely a fifteen-month hiatus, an unaccustomed period of unemployment for the homicide. Interrupted business affairs picked up their course again, and England was more a slave than ever to its sales counter.

I have tried to demonstrate it elsewhere, commercial abjection is inexpressible. It is the lowest rung and, in chivalrous times, even in England, mercantilism was dishonorable. What to think of an entire people who live, breathe, work, procreate for nothing but that; while other peoples, millions of human beings, suffer and die for great things? For ten years, from 1803 to 1813, the English paid to be able to traffic in safety on their isle, so as to cut the throat of France which went contrary to their villainy, Napoleon's France which they had never seen so great and which filled them with concern.

"Five hundred years of rivalry have made it personal, in each particular, the emulation that needled two peoples... France is in the position of ancient Rome relative to Carthage between the second and third Punic wars... England is the natural enemy of France; it is an avid, ambitious, unjust enemy of bad faith. The invariable and cherished object of its politics is, if not the destruction of France, its humiliation and ruin... that reason of State takes top priority over all other considerations, and when it speaks, all means are justified, legitimate, and even necessary provided they are efficacious." *Justa quibus necessaria.*[29] Thus

---

[29]*Justa quibus necessaria*: Latin for "Wars are just for those who find them necessary," Edmund Burke in his *Reflections on the Revolution in France*. The full phrase in Burke is "*Justa bella quibus necessaria.*"

they expressed themselves, the writers of civil law, prior to the Revolution.

But England was not only the natural enemy of France. It was its *supernatural* enemy. Three centuries earlier almost, – before the impure demons of Protestant mercantilism raged beneath the hateful Elizabeth's skirts, – the father of that crowned mare, the polygamous Henry VIII, had needed merely to gesture with his hand for all England, formerly called the Island of Saints, to turn its back on the Church. Major and initial shame of that realm devoted to Satan by a master kneaded out of mud, impatient with a religious authority that opposed its licentiousness. Instantaneously, *free* England, apostatic and all the more willing than its king, magnificently granted the property of bishops and monasteries to its obedient domestics. There were martyrs, but in small number. All that while France, convulsed with horror, fought with rage against heresy and prepared to combat it for fifty years by all means, until the abjuration, such as it was, of another ribald constrained to accept mass so as to reign over the spiritual progeniture of Saint Denys and Saint Martin.

While waiting for England to carry that iniquity unto the day of universal Judgment, while waiting also for the calamities that could be, one day, the very near consequences of it; there, at the time of Napoleon, the great insular anguish causing a Danube of blood to flow throughout Europe, and there was, above all, that horror of a cow with the four feet of the Golden Calf, rousing a mercenary continent for the destruction and abasement of the marvelous na-

tion of France! The darkest schemes of the most astute politic were its practices, and the very fear of revolting all civilized peoples did not stop it. It suffices
to recall the bombardment of Copenhagen, an incomparable act of piracy, the day after Tilsit, in order to
steal the Danish fleet which the English cabinet supposed belonged to the Franco-Russian alliance, no
hostile act having provoked that attack.

"The occult and magnetic power of England!"
Where exactly did I read those words? What was that
power and from where could it come to that apostatic
nation towards which turned, as towards a pole, the
needles of all the dirty or perturbed consciences, as
soon as the sorceress whispered in the silence of all
European chancelleries. Does that not seem too
frightening when one comes to think about it, that the
greatest of men was its victim and that the lion of the
desert that he was could be bewitched, in the end, by
that serpent of very low places, to the point of jump-
ing into its gob as if into a refuge!

It is mind-blowing to think that the man of
war to whom no other being can be compared was
vanquished by a Wellington! The truth is that at that
time his lieutenants were obeying him poorly and betrayed him. But, all the same, a Wellington, it is just
too ignominious! All that can be said about that inconceivable English general whose principal merit in
Spain was that of a good intendant of victuals and
who would have been infallibly crushed at Waterloo
if Napoleon had been able to make himself obeyed;
all that French indignation or sarcasm could inspire
would go no further to dishonor such a puppet than

the satirical counsels given to the "generals in chief"
by the English author of that charming work, *Advice
to the Officers of the British Army:*[30]

> "Nothing is so recommendable
> as generosity towards the enemy. To
> pursue him, giving him no quarter af-
> ter a victory, would be to take advan-
> tage of his distress. For you, it suffices
> to have proven that you can beat him
> when you deem it appropriate... You
> always act openly and in good faith to-
> wards your friends and enemies. In
> this way, you will guard against con-
> cealing a march or laying a trap. You
> will never attack the enemy during the
> night. You will remember Hector go-
> ing to battle with Ajax: 'Heaven, strike
> us by lightning, and fight against us!'
> If the enemy withdraws, let him gain a
> few days' head start, in order to show
> him that you do not expect to surprise
> him, when you undertake battle. Who
> knows if so generous a procedure will
> not engage him to stop? After he has
> stopped in a place of safety, you can
> then begin your pursuit of him with all
> your army... Never put forward an in-
> telligent officer; a good, fat companion
> is all you need to execute your orders.
> An officer who has an *iota* of knowl-
> edge above the ordinary, you should

---

[30]*Advice... Army*: published anonymously in 1782, but attributed
to Francis Grose.

> consider him as your personal enemy,
> for you can be sure he laughs at you
> and your maneuvers."

It is incontestable that Wellington, so justly admired by England, followed, to the letter, in his campaigns on the Peninsula and even in Belgium, these precious counsels. He had needed, in Spain and in Portugal, so as not to be destroyed twenty times, the essential absence of Napoleon and the criminal anarchy of the generals who stood in for him.

One can be quite certain that even the loss of the Empire was less bitter to Napoleon that that ridiculous and ignominious supplantation. What prevailed against him, the grandiose and magnificent Latin emperor, was, in the person of the mediocre Wellington, all the boutiques and all the strongboxes of London. It was the hideous hypocrisy of parsimonious and arrogant Protestantism by the discounters of carnage and infamy. It was finally and primarily the surprising subsannation of the God of armies repenting, as during the Deluge, for having made a man so great and, by the effect of a terrible mercy, humiliating him, in the end, under the feet of an abortion of glory!

Chapter 11: The Mercenaries

After England's shame, the shame of other European monarchies. One has to admit that this was disconcerting. Never a like prostitution had been seen. Catholic Austria, Lutheran Prussia, schismatic Russia, soliciting each in their turn or simultaneously English subsidies for the extermination of France. From '93 to 1813, five great coalitions, to say nothing of the innumerable and incessant subaltern plots whereby the world collected its money, powerful and haughty ministers, simple spies or scouts, united in the same design, waiting to devour each other when the common enemy was beaten. For twenty years, it was an inexpressible swarming of traitors, liars, assassins for hire, not stopping to put out their avid hands to England, who paid them while making a sour face and regaling them with the gratuity of its disdain when they had done a poor job, which happened extremely often. England's disdain! They had to swallow that at the same time as the appalling military penances inflicted on them by the invincible general.

Most certainly, it is constant tradition, the unchangeable jurisprudence of men of State, that all means are good in politics and that money even is ennobled by the intent to betray or murder. It is the doctrine of brigands, and Europe was a broad road. One got used to it, and the territorial dismemberment that followed Napoleon's fall, the *bartering*, according to the diplomatic usage of the time, established superabundantly the durability of those maxims.

At first, Austria. *Extra statum nocendi*, as Kaunitz had said in 1788. "Incapable of harm." Those instructions regarded Prussia at that time, before they became the universal watchword against France. *Harm* meant, to Austria, not being submissive and as France, immediately after the Revolution, harmed it in every way possible, Austria did not hesitate to backfill, by means of the classic English money, the disquieting deficit in its war chest. With the cynic Thugut, Choiseul's old spy, traitor of France and Austria, the haggling began, the dirty transactions. "We don't have a cent," it moaned as early as '94.

Metternich was bound to continue, but without the same frankness, being of a superior extraction and one of the most notorious gentlemen one could know. Napoleon having grown stronger, he went so far as to sell him very dearly an Archduchess, excellent business for the Austrian sovereign who was happy to sell his daughter, having passed the age of prostituting himself. It is difficult to imagine so perfect an abjection. When Napoleon's luck seemed on the verge of running dry, they remembered the Britannic coffers, and the father-in-law, Apostolic Majesty, armed three hundred thousand men so as to conquer an adulterous bed for his dear child, who found all that to be very good. They had, moreover, done all they could so that that happy change should prove inevitable. Long in advance, the Dominator's ruin had been decided by whatever means possible, and the marriage was merely an expedient to put him off his guard. It is in this way that the Prince of Metternich, writing later in his *Memoirs*, could apply this testimony to himself: "The views that have always formed the basis of Austrian politics are the *purest* that could be conceived."

With Prussia, there can be no question of any purity. It is nothing but boors and bandits with that bunch. "War," as Mirabeau said, "is the national industry of Prussia." One knows what that means. Since the barbarians of the Vth century, never has a more savagely brutal and pillaging nation been seen, and that has not changed. One can see the evidence of that in 1870.[31]

Its scandalous prosperity had begun, as everyone knows, in the XVIth century, with the union of the March of Brandenburg and Prussia such as it was, exiguous and very poor at that time, two German colonies in a Slavic state. That sad duchy of Prussia, formerly founded on the idolaters of the Teutonic Order, without frontiers or geographic delimitations, did not hesitate to become Lutheran to become stronger. It was the right thing to do in the XVIth century.

Teamed up by apostasy with the margravate of Brandenburg, it considered everything in its vicinity as good for the taking, and so it was, under the Hohenzollern, its unique reason of State. The great Frederick, veritable founder of Prussian power, took with both hands and as much as he could, swallowing Silesia and Poland, designating to his successors Saxony, Westphalia, Bavaria, Austria even if possible, all of Germany. But his immediate heirs would have need

---

[31] 1870: scil., the Franco-Prussian War.

of a kind of genius, the unfailing will of that redoubtable robber, and the oafish monarch, his grand-
nephew, wanting to oppose Napoleon, would certainly have lost everything if it was not for the deplorable
magnanimity of his adversary.

After Jena, Prussia having become poorer than
ever, it became expedient to practice a bit of prostitution, its temperament and its conscience not repug-
nant at all to the idea of it. England provided for it
with more abundance than love in 1813. Carthage
was forced to settle up with its mercenaries. Stein,
Scharnhorst, Gneisenau, and that horrible villain
Blücher served Prussia with a zeal all the more lively
since they intended to fatten up their dirty fatherland
with some of the best morsels of the beaten monster.
The most insatiable courts of Europe, considering
themselves the most damaged, maneuvered to have
the largest part in the present and future. The endemic
and hereditary brigandage was amplified, extravasat-
ed, magnified until it gave birth in our days to the
German Empire which will finish perhaps by gnawing away at itself, like those who are buried alive, in
the sepulcher of disdain and execration that socialism
is in the process of readying it for.

Must one register Russia among the mercenaries? Absolutely. One does not imagine Souvorof, for
example, traversing all of Europe, inundating Italy,
and grimping up Swiss mountains, short of money.
The Muscovite accounting was uncertain, and the
Russian currency probably depreciated on the other
side of the Vistula. One also does not imagine the de-
licious parricide Alexander pulling himself away

from his little pleasures in order to go and have himself massacred at Austerlitz or at Friedland. The role of an Olympian negotiator suited him better and was less of a burden on him.

Of all Napoleon's mistakes, after that of Bayonne, the gravest and most difficult to expiate was that of letting himself be taken in by the smiles and caresses of that Byzantine who did not let a single day pass in which he did not betray him, whose fully enthusiastic friendship was a Greek lie imperturbably sustained for four years until England, grown impatient with that romance, constrained him to declare what he was in reality: an implacable enemy.

When the coalition of 1805 was fully baked and ready to serve, England had concluded a treaty of subsidies at the rate of 1,200,000 pound sterling per one hundred thousand men whom Russia would equip with arms, thirty million francs for Austerlitz. The treaty did not say whether the dead would be deducted. Did the crestfallen Tsar obtain quittance after having experienced in Moravia that it was less easy to win a great battle than to assassinate his father? The fallen angels ought to know, but that is infinitely doubtful for men. Business is business, and dissatisfied England held onto its cedula. The Russians weren't being paid to be beaten. That account was doubtless settled by infractions of the continental Blockade.

On the day after Austerlitz, Alexander, whom Napoleon could have kept as a prisoner of war and locked up in a fortress, very humbly supplicated his conqueror to allow him to withdraw with the rest of his army, which request was granted. "To grant them mercy today," exclaimed the heroic and miserable Vandamme, "means they will be in Paris in six years!" Ten years later, on Saint Helena, a pensioned representative of Alexander's was there to make sure of the captive's detention. Such is the beauty of history, such is politics, and such was the recompense for the magnanimity of Napoleon, who almost always forgave and was never forgiven.

It remains to be seen what became of his soul, his too great soul, in that horrible whirlwind of iniquities. Soul of a sublime high-school student, carried by God's Breath to unknown heights, no longer seeing human pettiness almost, incorrigibly in love with everything that appeared to him to possess generosity or grandeur and, because of that, despite the most sumptuous genius, marked, much more than an ordinary soul, for all the sufferings of Deception.

There is, in the humblest churches of France, a poor lamp lit night and day, before the Holy Sacrament of the Altar. The thought crossed my mind, absurd perhaps, that that lamp is something like Napoleon's confidence.

Chapter 12: The Greats

When Napoleon restored the rank of marshal, giving himself in that way eighteen *cousins*, he appeared to be afraid of his action. That remark is by the contemporary Thiébault, admirably situated to pass judgment, whose *Memoirs*, much superior to Marbot's, are, mainly from the military point of view, the most faithful document that could be consulted.

Afraid, he supposes, of handing too much power to his old companions of war, having become his subalterns and his subjects, the parvenu Emperor estimated that, to a certain degree, it would be more suitable to "restore that institution with his choices, and he made them in such a way that the favor part of it entirely dominated the meritorious part." Naturally, I leave to Thiébault the responsibility of so grave an accusation, by remarking, nonetheless, that it is quite troubling to see Napoleon place at the same level of supreme honor generals whom he must have known, better than anyone, the inequality of.

A certain Berthier, for example, called by the emperor himself a "gosling", or a pseudo-conqueror like Brune, next to the great Masséna; the heroes Ney and Lannes, comparable only to chevaliers of ancient times, next to a Soult, invisible and nowhere to be found at Austerlitz as long as any danger remained, where his army corps had the principle role, and who, later, attributed all the glory to himself. That duke of Dalmatia to whom Napoleon did not want to attribute the name of any place that memorialized a victory was, as one knows, the most effective artisan of the insuccess in Spain where the emperor had had, after the quasi betrayal of Oporto, the weakness or inconceivable blindness to entrust him with a preponderant situation.

But what to say about Marmont, the vanquished of Arapiles and the abominable traitor of Essonne whose name alone became a bloody curse? What to say about Murat and Augereau, so intrepid however the both of them, and who were so horribly unfaithful on bad days. What to think about the imbecile and vain MacDonald, pillager of Italy in '99, who never knew how to do anything but fight; or Gouvion Saint-Cyr, perhaps the most able general there was in all of Europe after Napoleon, but whom a fit of diabolical humor made lose the advantage in the battle of Dresden and began the irreparable disaster of 1813; or the inept and valorous Oudinot; or the ridiculous drummer Victor, canonized the Duke of Bellune; or the ferocious and set-in-his-ways Davout depriving invaded France of an army that maybe would have saved it, while doggedly persisting, with the obstinacy of a brute, to defend a city that no one was attacking; or Grouchy finally whom the guardian demon of England seems to have designated as the unfortunate emperor's choice for bringing his pilgrimage to a close.

The most incomprehensible and most funereal among those insensate promotions was certainly that of Bernadotte whom Napoleon knew to be his personal enemy and whose military boastings he must not have esteemed all that much. One knows how he was repaid for it. But Bernadotte had in his favor his being Joseph's brother-in-law, and Napoleon was the leader of that very sensitive clan. That family tie got him pardoned, after several other things, for the crime of Auerstadt which another prince would have made him pay for with his head, and his very strange conduct at Wagram earned him merely a benign and temporary disgrace. Having become the king of Sweden with the consent of his Master, who didn't have the character to oppose it, that odious adventurer, intoxicated to see himself "praised by legitimates," soon became the fierce enemy of his benefactor and his fatherland. His name is a piece of garbage in history, and it is perfectly suitable that all the Lutheran renegades amongst the Swedish should be proud of and satisfied with him.

Such was, nearly always, Napoleon's gain when he wanted to make the men surrounding him great. A son of the Revolution, he needed to take what his mother had given him naturally, that is to say villains or domestics 90 or 95% of the time. Of servants possessing great talent, those outside the military, to name only Talleyrand and Fouché, they were, under him, the marvelous scoundrels that they would have been under any regime. One can even say, without hyperbole, that their turpitude suffered the contagion of his grandeur, to the point that the world will perish clearly before one is able to discern for them a sufficiently equitable contempt. And they were almost all like that, as appropriate, at each of the infinite levels of administration of the Grand Empire, so that one ends up by being less surprised by Napoleon's glory than by the ignominy of ingratitudes or betrayals that his reign had determined, the excessive energy of the star having activated universal putrefaction in an unprecedented manner. When it waned, there was an unknown stink...

Truth be told, Napoleon never knew how to punish adequately, and that is seen at every instant, it is found on every page of his life, until one grows impatient with him, it is perhaps the essential trait of that strange man among strange men whom one has so wanted to represent as a tyrant and who was principally, by virtue of one does not know what heredity, a profound fatalist, incapable of resentment, always afraid of destroying something of his work by lowering those whom he had elevated, ceasing to want and ceasing to act when he thought he had heard the voice of destiny – sitting down then, full of a mute resignation, on the edge of the well of despair.

"Complaining," he said, "is beneath my dignity and my character. I command and I keep quiet."

Chapter 13: The Sacrifices

How many were they, those there? Five or six hundred thousand maybe. One does not know. That can go as high as a million of French victims, not by the ambition of their leader as has so often been said, but by the strength of things that was none other than Divine Will.

No one in Europe desired peace as passionately as Napoleon did, because he needed peace to institute the magnificences that his marvelous mind had conceived, and he could never obtain it. From '96 to 1815, he fought for the hugely desired conquest of that terrestrial Paradise before the gates of which all his armies came to be crushed.

And what armies! Never has anything so beautiful been seen. To engender and to produce, in the end, those armies, dreamt of and apotheosized, – the painful gestation of forty centuries had been needed. At first were needed the poor and sublime Bishops of barbarous Chaos and all the Saints from the Time of the Merovingians or Carolingians who had amalgamated the land of France with the very precious Blood of Christ; the Chivalry of the Crusades had been needed then and its supernatural enthusiasm; and then the horrible tribulation of the Hundred Years' War with the English, the appalling convulsions of the XIVth and XVth centuries when the Kingdom of the Mother of God thought it was going to die; the dungheap of all the Bourbons then, finally, was needed, and all the guillotines of the Terror. One does not know a nation that has been so belabored, so *amended* by blood and filth.

Impious, – certainly it was or appeared to have become, – like all the world, moreover, even in Spain, and there was hardly anything else one could be at the end of the XVIIIth century. But that was, in France, a superficial impiety, a thin layer, a sort of spiritual scabies contracted under the Bourbons, able to be cured by baths of blood or fire, and not involving the entrails. France is simply incurable of God, the most diabolic experiences have demonstrated it, that of the Revolution above all. Precisely because it was the most generous of nations, it was impossible that, deprived temporarily of the Christian faith, it did not rush headlong into the magnificent deception of '89 and the horrifying deliriums that followed by consequence. Because what was necessary for that abandoned Visitandine was a representative and bodily God, a tangible God that might console her, when Napoleon was shown her, – she *recognized* him immediately, she let out an immense cry of boundless love, and she abandoned herself entirely.

Whether it is a question of Fréjus or Golfe-Juan, there is not in all of history another example of so prodigious an ascendant. That extraordinary man was really God for his soldiers, who were the flower of France. He could do anything he wanted with them, his exorbitant soul absorbing, as I have said, all those souls having become his own because he willed it, perhaps also without his having willed it, for all that is really very mysterious.

That armed people followed him everywhere,

accepting, for the love of him, all the difficulties of life and all the torments of death. When the greats whom he showered with his benefits betrayed him, the poor soldiers who had conquered all the earth, rich only in their wounds and their glory, remained faithful to their fallen Emperor, to their captive and deceased Emperor, not succeeding in understanding that it was over forever. Villages in all the provinces saw die, more than sixty years ago, those invalid and poverty-stricken, naïve and grandiose orphans of the Prodigy, who always saw themselves in Egypt or at Moscow. With them, the stars seemed to fade.

Their memory diminished, and a new generation of men, who were able to catch a glimpse of them merely in the legendary images of Charlet or Raffet, were ignorant of them in reality, uncertain whether like men could have existed as the companions of the Giant whose name alone belittled all their grandeurs.

A day will come maybe when Napoleon's relics will no longer be resting in his admirable Tombeau des Invalides. The coffin will be opened and it will be *empty*, the very appearance of that dust unable to exist any longer after the extinction of the prestige that surrounded him. It will have been reunited with the confused and dispersed dust of the humble soldiers who sacrificed themselves for their leader, and whose loving children's souls will be gathered around His, on the day of Universal Judgment, just as his invincible Guard did, formerly, on days of great battle!

Chapter 14: The Guard Falls Back!...

One will be incapable of understanding anything about Napoleon as long as one does not see in him a poet, an incomparable poet in action. His poem is his entire life, and there is nothing that equals it. He always thought as a poet and could act only as he thought, the visible world being for him merely a mirage. His astounding proclamations, his infinite correspondence, his visions on Saint Helena bear sufficient witness to it. When he spoke, when he wrote, his language magnified everything.

One does not tire of re-reading his admirable letter of February 2, 1808, to the astute parricide Alexander, extremely indignant on reception of it and certainly incapable of understanding it. He offered him no less than to share the world, showing Asia to him and reserving for himself all the West; that, not like a magnificent eventuality, but like a necessary consequence of their system of alliance: "... Then the English will be crushed under the weight of events whose atmosphere will be charged. Your Majesty and I, we will have preferred the sweetness of peace and of passing our life in the midst of our vast empires, occupied with invigorating them and making them happy... The enemies of the world want nothing to do with it. One must be greater, in spite of them. It is wise and politic to do what destiny commands and to go where the irresistible march of events leads us."

Always the destiny! Napoleon is he then the poet of destiny? The events he speaks of have demonstrated historically the irreality, or, if one prefers, the inanity of his great designs, but they did not demonstrate it in the soul of that Emperor of emperors, where they had, clearly, a prophetic consistency, a indemonstrable reality, all the more certain in his eyes. Discerning better than anyone the material appearances in war or in the administration of his empire, he had, at the same time, something of an ecstatic presentiment of what was expressed by those perishable contingencies, and it is precisely that which constituted the poet in him.

It was not possible that his sentimental life should differ essentially from his public life. That disparateness can belong only to ordinary great men, to the rabble of great men. Napoleon owed it to himself to act, in love, in the same way that he acted as emperor, in other words, like an extremely great poet, indiscourageable poet of marvelous illusions which were sufficient unto him in the gorgeous crepuscule of a summer morning that was his entire life. The greatest disasters and his dreadful fall even did not succeed in waking him up at all. On Saint Helena, he continued his dream while suffering, and since his death he continues it still in the imagination or in the heart of those who admire him.

It has been said very exactly that Napoleon loved like a schoolboy. When would he have been able to find the time and the experience to love in any other way? Like all schoolboys he loved prostitutes, women giving themselves on a dime, with or without fuss. One can even say that having been responsible, at a very early age, for the affairs of the entire world, he didn't have the leisure to love or to marry others and that, a little later, it must have seemed unimportant to him. His passion for Josephine who was and who remained a hussy, passion attested by letters filled with frenzy, has the exact same character of sensual transport as that of an imaginative adolescent, still a virgin, set on fire by the coquetry of an ambitious woman.

That kind of eruption, to speak decently, is easily curable and the schoolboy does not take long to be instructed. Moreover, at the beginning of his passion for Josephine, his future greatness was only guessed at or anticipated. The perverse and fascinating creole was his first smiting. He who was still referred to only as Bonaparte and who, later, would have merely had to wave his hand and the loftiest virtues would have immolated themselves for him, must have thought then that an Olympian goddess was deigning to condescend, far beneath herself, to him: "*Mio dolce amor*,[32] do not give me kisses, for they set my blood on fire." One writes things like that at eighteen years old. But it appears that in love Napoleon was always that same age. Three lusters after his great passion for Josephine, there was Marie-Louise and the surprising childishness of his equipage at Compiègne. It is important to remember that the doll that had been sent to him from Vienna was the daughter of Caesars and that that gave him a new smiting which renewed in him the schoolboy tenacity

---

[32]*Mio dolce amor.* Italian for "my sweet love" or "love of my life."

of his earlier days.

Those two women, very dignified the two of them, were both equally unfaithful and traitresses, as appropriate. A fatalist, as I mentioned above, he got along as best he could, having enough to do to rouse against his single person all the European peoples in order to fulfill what he called his destiny. The epoch, moreover, wanted it like that. Each did whatever one wanted, and Napoleon's sisters were plainly courtesans. Caroline, the most odious of the three, not content with having dishonored her husband twenty times, the unfortunate Murat, made that hero of all battles into a lamentable victim of hallucinations, whom she believed she had need of in order to assassinate her brother.

But the immense poet of the twenty-year Epic, – who could assassinate him or merely aggrieve him in mortal fashion? He saw his wives, his sisters, his brothers armed against him, just as he saw his ungrateful lieutenants, just as he saw all things: in the enigmatic mirror of his magnificent thought.

He had what is called, in common parlance, mistresses, as great a number of them as he wanted, and in passing, being the soldier among soldiers in the world, but they did not possess and did not know his soul. "*Ubi thesaurus, ibi cor*. Where the treasure is, there is your heart." Napoleon's heart was not an impregnable citadel, but those men or women who penetrated it believed there was nothing there, because the treasure was *invisible*. That treasure was the secret of his grandiose poetry, the arcana of that Prometheus ignorant of himself, whose gravest faults had that same excuse that Polyphemus or Antaeus did, that he did not know himself to be so colossal or so predestined. That was, with his impatience of all obstacles, the profound zeal of a supernatural mission that he never came to untangle, but which exuded from all his pores and the certitude of which crucified him; – an *amorous* situation that demonstrated him to be, all the same and forever, infinitely above ordinary desires and their miserable servitudes.

I have mentioned Napoleon's two smitings. There was a third, even more fatal. That was the smiting of Defeat. Up until Waterloo, he had known disasters, but he had not known defeat. That other prostitute, so long excluded, wanted him as well, and he had to submit finally.

The Guard falls back!... With this panic cry, he sees his battle line crumble, he sees his last army put to rout, he feels the grip of the monster, and his virginity as a conqueror is lost. An awful darkness falls and surrounds his soul. Is it all completely over then? Must the poem end on that dreadful note? Where now his star? What happened to his heart and his treasure? Clearly it is not Wellington who stole them, and it is not that Prussian boor who did. He will discover it again in three months, at two thousand leagues from his capital, in another hemisphere. But then his star will be like a poor woman asking for bread, his heart will be tortured and his treasure will be sorrows. Ah! it is not just the Guard that falls back at Waterloo, it is the Beauty of that poor world, it is the Glory, it is the Honor even; it is God's and men's France having become a widow all of a sudden, going

off to weep in solitude after having been the Dominatrix of nations!

The morning of that terrible day, the Church
Militant celebrated, in all the parishes of Christianity,
the mass of two very ancient martyrs and reminded
all its faithful to "take glory in tribulations, *gloriamur
in tribulationibus.*" There were certainly, in France,
humble priests and even humbler assistants who remembered then their close friends and relatives who
had gone into battle and who did not think any more
than their Leader to invoke those old martyrs. It is
probable however that many of those who were immolated were assisted by them in their agony; but the
gentle and mystical murmur of that prayer had no other appreciable echo than Cambronne's desperate imprecation,[33] and the beaten Emperor hardly thought to
glory in his torment.

He gloried in it later, on Saint Helena, when
he saw coming towards him the great Lover of beggars and emperors, and she took away with her his
Secret, which was not to be conveyed to anyone.

---

[33]Cambronne's imprecation: Pierre Jacques Étienne Cambronne
(AD 1770-1842), a French general under Napoleon, and present
at the Battle of Waterloo, is said to have said "Merde!" ("Shit!" or
"Hell!") when asked to surrender. He later denied saying such a
thing.

Chapter 15: The Invisible Companion

It is taught that each man is accompanied, from birth to death, by an Invisible thing, charged with watching over, attentively, his soul and his body. That Invisible thing is called his Guardian Angel, requisite protector of God capable of belonging to one or another of the Nine Angelic Choirs.

That is the universal belief of Christians. That perpetual companion is at one and the same time an inspirator and a judge. Lofty thoughts ensue from it, and what one calls reproaches of the conscience, it is the guardian angel that makes them be heard. It knows what we do not know, it sees what we do not see, it is always present in us and around us, inexpressibly respectful of our freedom, knowing the real greatness of our souls and the inconceivable dignity of our bodies of dirt and sediment called on to shine, when we have finally woken up. When a man does evil, the angel withdraws silently into the deep places of the criminal soul where the sinner himself does not penetrate, and it weeps as only Angels know how to weep.

"... If life is a feast, they are our guests; if it is a comedy, they are our associates, and such are the formidable Visitors during our sleep, if it is nothing but a dream!... They are our very close friends, the perpetual Voyagers of the luminous Ladder of the Patriarch, and we are informed that each one of us is jealously guarded by one of them, like an inestimable treasure, against the devastations of the other Abyss, – which gives a most astounding idea of human nature.

"The most sordid scoundrel is so precious that he has, to watch exclusively over his person, someone similar to Him who preceded the camp of Israel in the column of clouds and in the column of fire, and the Seraphim who burns the lips of the most immense of all the prophets is perhaps the conveyor, as large as all worlds, charged with escorting the very ignoble cargo of an old soul of a pedagogue or magistrate.

"An angel recomforts Eli during his famous fright; another accompanies the Hebrew Children into their furnace; a third shuts the gob of Daniel's lions; a fourth finally, who is called 'the Great Prince,' disputing with the Devil, does not yet find himself colossal enough to curse him, and the Holy Ghost is represented like the only mirror wherein those unimaginable acolytes of man can possess the desire to contemplate themselves.

"Who are we then, in *reality*, such that these defenders might be appointed to us and, above all, who are they, those beings enchained to our destiny and whom *it is not said* that God made like us, in his image, and who have neither a body nor a face? It is about them that it was written never to 'forget hospitality,' for fear that one of them might be hiding among necessitous strangers."[34]

Who then could have been *stranger*, more *ne-*

---

[34]Original footnote: LÉON BLOY, *La Femme pauvre*.

cessitous, than Napoleon? Understanding nothing about the appearance of such a man on earth, I renounce describing it, and how could I speak of Him who was charged with accompanying him invisibly everywhere? One would be compelled to attribute to him a Cherub, a Throne, a Domination, or at least a very great and very splendid Archangel. I think, on the other hand, that he had to have for a guardian one of the least spirits from the lowest rung of the celestial Hierarchy.

A mediocre Judas such as Bernadotte, for example, could have need of being attended to by one of the loftiest princes or ministers of Grace, capable of carrying the mountain of his betrayals and removing from him, – horribly – all human punishments, while waiting for the hour of God and his Justice. But it could not be like that for Napoleon. What that extraordinary personage needed was the guardian angel of a little child abandoned on the footpaths of the world, a modest protector to keep away all vagabond dogs, to guide his way among the bramble bushes and stones that could have offended him, a humble and quasi timid guardian angel for the greatest of all men! A very sweet and invisible friend, deferent and grave, to say to him from the bottom of its heart:

"Forgive often, but do not always forgive. God made you the father of fifty million of his creatures who cannot know who you are, given you do not know yourself. Do not devour those unfortunate people who are in the Image of God and in your own image. You are allowed to enchain kings and trample all over them because they are abhorred by the Holy

Ghost *whom you signify perhaps*. Only, do not be too clever and do not undertake to do away with the mountains that belong to God. Until then, you will be invincible, but not much longer, and you will understand soon enough. The snow and deluge are on their heights; do not force them to come down."

What astonishing colloquies those two Imperturbables had, the one of the earth and the other of heaven, the one visible and the other invisible! And Napoleon, he too, was he not invisible in his own way, and how very much! to his servants who were incapable of suspecting or even supposing his anxieties, when he conversed with the translucent Companion through whom his anguished soul saw tempests forming. "Do not go there," said the angel. "My destiny commands it," said the emperor. And behold Destiny pitting itself against God, and then Napoleon was lost! But that there, nobody in his entourage could see it. There were moments like that, hours, long nights, when that Master of the world, not knowing what to do, passed from one resolution to another, stepping over the reefs so as not to be immediately led back with violence by the insulting waves until, worn out by the effort, he let himself fall with five or six hundred thousand men, while murmuring one does not know what words, the equivalent perhaps of this: "God have pity on me!"

That wreck of human majesty, nearly infinite, arrived finally on Saint Helena. At his disembarking on that island, which because of him has become famous ever since, the admiral Cockburn had arranged that he be given an invitation addressed to "general

Bonaparte." On receiving it from Bertrand's hands, Napoleon said to the great marshal: "You must send that to general Bonaparte; the last time I heard mention of him, it was at the battle of the Pyramids or that of Mount Tabor." Lord Rosebery, veritable Englishman for all that, calls it an indignant and revolting buffoonery, that obstinate refusal on the part of the English to employ the imperial title belonging to the great Captive.

The same Cockburn responded in the following terms to a letter wherein the count Bertrand mentioned the name of the Emperor: "Sir, I have the honor of informing you of my reception of your letter dated yesterday. That letter obliges me to declare to you officially that I have no knowledge at all of an emperor of any sort residing on this island, nor any person assuming that dignity having, as you mentioned, voyaged with me on the *Northumberland*."

That low and ignoble English persecution lasted for a longer time than Napoleon himself. "On the coffin of the Emperor," said Rosebery, "his servants wanted to write this simple word: *Napoleon*, with the place and date of his birth and death. Sir Hudson Lowe refused to consent, unless one added also the name Bonaparte. But the servants could not accept a designation that the Emperor never wanted to admit to. So the coffin bears no name at all. That seems incredible, but that's how it is."

Nothing was spared in the punishment of him whose unpardonable crime had been infinitely to surpass all human heads and to have accomplished the greatest things ever seen on earth, for nineteen centuries. Nothing except the victim's moaning and perhaps also his *presence*. The executioners and English domestics were clearly right, more than they knew, to deny the presence of the Emperor Napoleon. He possessed merely a small human appearance already, having been touched by death. Napoleon was out of their reach, as much as his invisible Companion was, whom he was conversing with, far from them.

One has often spoken of his continual monologues, so often interrupted by the objections he made to himself. In reality, his monologues were the dialogs of an Absent being with an Invisible being, and that latter being being none other than the comrade that was needed, in that excessive misery, by an exile who was no longer able to obtain even that he be addressed by his proper name.

One can suppose that at that last hour, a powerful Archangel must have intervened to present to the Father of mercy his greatest image, but, during the course of his journey of glory and misfortune, it seems conformant with the laws of supernatural equilibrium that the Emperor of the centuries should have had, for protector and companion at every moment, the least of the blessed Messenger Spirits that the Lord could find in his vast heavens.

– Bourg-la-Reine, January-April, 1912.

Other Books by the Publisher

Fanchette's Pretty Little Foot
by Restif de La Bretonne

Je M'Accuse...
by Léon Bloy

My Hospitals & My Prisons
by Paul Verlaine

Salvation Through the Jews
by Léon Bloy

Words of a Demolitions Contractor
by Léon Bloy

Cellulely
by Paul Verlaine

Flowers of Bitumen
by Émile Goudeau

Songs for Her & Odes in Her Honor
by Paul Verlaine

On Huysmans' Tomb
by Léon Bloy

Ten Years a Bohemian
by Émile Goudeau